Dynamic Combinatorial Chemistry

Dynamic Combinatorial Chemistry
In Drug Discovery, Bioorganic Chemistry, and Materials Science

Edited by

Benjamin L. Miller
University of Rochester
Rochester, New York

A JOHN WILEY & SONS, INC., PUBLICATION

For general information on our other products and services or for technical support, please
contact our Customer Care Department within the United States at (800) 762-2974, outside the
United States at (317) 572-3993 or fax (317) 572-4002.

Wiley also publishes its books in a variety of electronic formats. Some content that appears in print
may not be available in electronic formats. For more information about Wiley products, visit our
web site at www.wiley.com.

Library of Congress Cataloging-in-Publication Data:

Miller, Benjamin L.
   Dynamic combinatorial chemistry : in drug discovery, bioorganic chemistry, and
      materials science / Benjamin L. Miller.
         p. cm.
   Includes index.
   ISBN 978-0-470-09603-1 (cloth)
1. Combinatorial chemistry. I. Title.
   QD262.M64 2009
   615'.19—dc22

                                                                                      2009019344

Printed in the United States of America

10 9 8 7 6 5 4 3 2 1

Contents

Preface

In a relatively short period, dynamic combinatorial chemistry has grown from proof-of-concept experiments in a few isolated labs to a broad conceptual framework, finding application to an exceptional range of problems in molecular recognition, lead compound identification, catalyst design, nanotechnology, polymer science, and others. This book brings together experts in many of these areas, as well as in the analytical techniques necessary for the execution of a successful DCC experiment. While there have been several outstanding general reviews of the field published over the past few years, the time seemed ripe for an overview in book form.

DCC is useful both because of its ability to rapidly provide access to libraries of compounds in a resource-conserving fashion (i.e., there are few things simpler than mixing molecular components and allowing them to "evolve" towards an optimized result), and because it can also yield completely unexpected structures, or molecules not readily accessible by traditional synthesis. As the reader will see, this book is full of examples showcasing both of these strengths. Challenges inherent in the DCC technique (or suite of techniques) and opportunities for advancement are highlighted as well, and hopefully will spark the development of new solutions and strategies. In some cases, particular examples are discussed in more than one chapter, in order to allow their exploration in different contexts.

The chapters contained herein cover the literature from the beginning of what came to be known as dynamic combinatorial chemistry (but was initially known as a confusing mix of things!) up to late 2008. A brief overview of historical antecedents to DCC is also provided. Of course, it is inevitable that despite the best of intentions, there may be research groups active in the field whose work is not covered as comprehensively as one might wish. We hope that any researchers thus inadvertently neglected will accept our apologies.

My personal thanks goes to the broad community of scientists working on DCC and affiliated techniques; I have been continuously pleased by your openness and helpfulness, and astounded by your creativity. Hopefully this book does justice to all of your efforts. Closer to home, the DCC projects that have unfolded in our group at Rochester occurred only because of the persistence and intelligence of my coworkers, and therefore I would like to thank Bryan Klekota, Mark Hammond, Charles Karan, Brian McNaughton, Peter Gareiss, and Prakash Palde for their efforts and continuing interest. Finally, thanks are also owed to Jonathan Rose, our editor at Wiley, for his exceptional patience during the process of assembling this book.

I hope you will find this volume to be a useful guide to the state of the art in DCC, as well as a source of inspiration for your own efforts in this field.

BENJAMIN L. MILLER

Rochester, New York
September 2009

Contributors

Marcus Angelin, Department of Chemistry, KTH—Royal Institute of Technology, Stockholm, Sweden

Jennifer J. Becker, U.S. Army Research Office, Research Triangle Park, North Carolina

Venugopal T. Bhat, School of Chemistry, University of Edinburgh, Edinburgh, United Kingdom

Michel R. Gagné, Department of Chemistry, University of North Carolina, Chapel Hill, North Carolina

Peter C. Gareiss, Department of Dermatology, University of Rochester, Rochester, New York

Soumyadip Ghosh, Department of Chemistry and Biochemistry, University of Maryland, College Park, Maryland

Michael F. Greaney, School of Chemistry University of Edinburgh, Edinburgh, United Kingdom

Lyle Isaacs, Department of Chemistry and Biochemistry, University of Maryland, College Park, Maryland

Rikard Larsson, Department of Chemistry, KTH—Royal Institute of Technology, Stockholm, Sweden

Takeshi Maeda, Institute for Materials Chemistry and Engineering Kyushu University, Fukuoka, Japan

Benjamin L. Miller, Department of Dermatology, University of Rochester, Rochester, New York

Hideyuki Otsuka, Institute for Materials Chemistry and Engineering, Kyushu University, Fukuoka, Japan

Sally-Ann Poulsen, Eskitis Institute for Cell and Molecular Therapies, Griffith University, Queensland, Australia

Olof Ramström, Department of Chemistry, KTH—Royal Institute of Technology, Stockholm, Sweden

Morakot Sakulsombat, Department of Chemistry, KTH—Royal Institute of Technology, Stockholm, Sweden

Atsushi Takahara, Institute of Materials Chemistry and Engineering, Kyushu University, Fukuoka, Japan

Pornrapee Vongvilai, Department of Chemistry, KTH—Royal Institute of Technology, Stockholm, Sweden

Hoan Vu, Eskitis Institute for Cell and Molecular Therapies, Griffith University, Queensland, Australia

Chapter 1

Dynamic Combinatorial Chemistry: An Introduction

Benjamin L. Miller

Darwin was the first to recognize (or at least the first to publish) the observation that nature employs an incredible strategy for the development and optimization of biological entities with a dizzying array of traits. From the macroscopic (i.e., giraffes with long necks) to the molecular (i.e., enzymes with exquisitely well-defined substrate specificity) level, nature generates populations of molecules (or giraffes) and tests them for fitness against a particular selection scheme. Those that make it through the selection process are rewarded with the ability to successfully reproduce (amplification), generating new populations that undergo essentially open-ended cycles of selection and amplification.

In the laboratory, biologists have directly benefited from the ability to co-opt Darwinian evolution: the polymerase chain reaction (PCR) [1] and the Systematic Evolution of Ligands by Exponential Enrichment (SELEX) [2,3] process are obvious examples of the selection and amplification of nucleic acids (and there are many others). Protein- or peptide-targeted approaches are also now commonplace: phage display, for example, has become a standard method [4]. In contrast, until recently chemists have had no such "evolutionary" advantage: while combinatorial chemistry brought about the advent of the synthesis and screening of libraries (populations) of compounds (or, more precisely, the *intentional* synthesis and screening of libraries since the properties of mixtures of compounds had been evaluated through the centuries as part of natural products chemistry, or inadvertently through the synthesis of mixtures), such methods are only a single cycle through the evolutionary process.

Dynamic Combinatorial Chemistry, edited by Benjamin L. Miller
Copyright © 2010 John Wiley & Sons, Inc.

No amplification step occurs, and the next step requires intervention by the chemist, in the form of synthesizing a new set of compounds (part of what we view as traditional medicinal chemistry).

Dynamic combinatorial chemistry (DCC) arose out of chemists' desire to couple selection and amplification steps to library production. In essence, DCC relies on the generation of a library of compounds under reversible conditions, and allowing that library to undergo selection based on some desired property. We discuss the components of this process in greater detail later in this chapter (and throughout the rest of the book). However, DCC built on a number of different lines of investigation, and it is useful to first discuss a few of these DCC antecedents in order to understand the intellectual foundations of the field. Perhaps the first recognition of a binding-induced selection process was that of Pasteur, who noted that crystals of tartaric acid could be sorted into mirror-image forms. Although the analytical technique here was certainly something one would not want to extend to large libraries (Pasteur sorted crystals by hand!), enantiomeric selection based on optimizing crystal-packing forces nonetheless demonstrated one component of the DCC process.

Many more recent experiments arose out of the body of researchers studying the molecular origins of life. Two areas of particular interest have been the origins of chirality and replication. Building on work by Miller and Orgel [5], Joyce et al. demonstrated in 1984 that diasteriomerically pure nucleotides would assemble on a complementary nucleic acid strand efficiently, but the presence of nonchirally pure materials would dramatically inhibit the assembly process [6]. An important DCC precursor—and evolution of Joyce and Orgel's studies—was reported by Goodwin and Lynno in 1992 [7]. This work demonstrated that trinucleotides bearing either a 5'-amino group or a 3'-aldehyde could be induced to assemble reversibly on a DNA template via formation of an imine. Subsequent work published in 1997 incorporated imine reduction into the process [8], effectively allowing single-stranded DNA to be used as a catalyst for the production of a DNA-like secondary amine. A somewhat more complex variation of the Pasteur experiment involves spontaneous resolution under racemizing conditions (SRURC) of systems such as bromofluoro-1,4-benzodiazepinooxazole, shown in Fig. 1.1 [9]. Crystallization of this compound from a racemic, rapidly equilibrating methanolic solution can lead to amplification of either enantiomer via the production of single-enantiomer crystals.

Product templating and re-equilibration of product mixtures have also been studied extensively in the molecular recognition community. For example, Gutsche and coworkers examined the base-mediated production

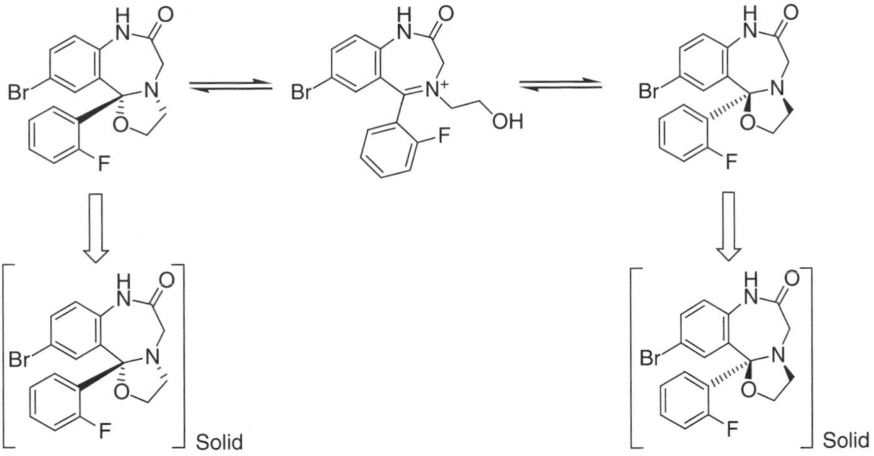

Figure 1.1 SRURC of a bromofluoro-1,3-benzodiazepinooxazole.

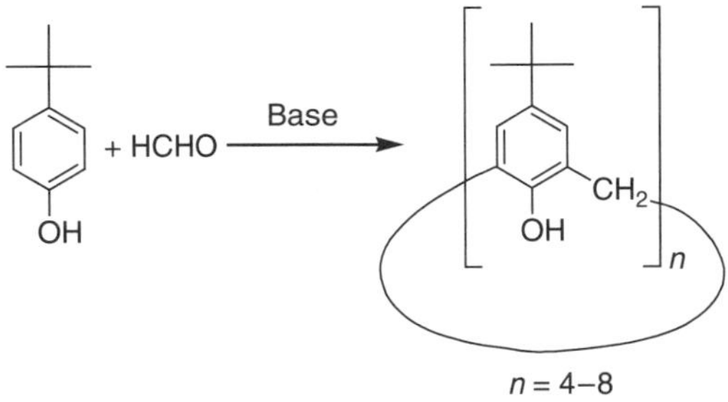

$n = 4-8$

Figure 1.2 Base-mediated synthesis of calixarenes.

of calixarenes from *para*-alkylphenols and formaldehyde (Fig. 1.2), and observed that product distributions were altered based on a large number of factors [10]. Of particular interest to DCC, the authors described calix[4]arenes as arising via a thermodynamically controlled process, in part via ring contraction of calix[8]arenes and calix[6]arenes. Thus, this may be regarded as an example of a dynamic self-selection process.

Molecular recognition is obviously a critical component (and often the primary goal) of DCC-based molecular discovery, and the molecular recognition community was instrumental in developing experiments that directly prefigured the development of DCC. Two examples from the Lehn group are illustrative. In 1990, Lehn and coworkers reported that mixtures of tartrate-based compounds could be induced to form liquid crystalline phases [11]. This recognition-driven supramolecular assembly was hypothesized to occur via formation of a triple-helix structure, mediated by nucleic acid-like hydrogen-bonding interactions. Three years later, the same group reported a particularly spectacular example of recognition-mediated self-sorting (Fig. 1.3) [12]. On treating an equimolar mixture of **1**, **2**, **3**, and **4** with excess $[Cu(CH_3CN)_4]BF_4$ in CD_3CN, a highly complex ^1H NMR spectrum was initially observed. This was found to gradually resolve itself into a spectrum dominated by the presence of the self-selected complexes **5**, **6**, **7**, and **8** (although small amounts of other complexes remained). Self-sorting among ligands predisposed to bind different metals was also observed when **9** and **10** were mixed with copper and nickel salts. Again, the authors initially observed production of a highly complex mixture, which resolved over time to consist primarily of copper complex **11** and nickel complex **12**.

With these selected examples as context, it became clear to several laboratories in the mid-1990s that one should be able to combine reversible formation of compounds (exchange processes) and a selection method with the then rapidly developing field of combinatorial chemistry to produce equilibrating libraries that would evolve based on some selection process. Thus, dynamic combinatorial chemistry or DCC, as it came to be called,[1] evolved from a number of lines of research into the diverse and vibrant field it is today.

1.1. The Components of a Dynamic Combinatorial Library Experiment

The design of any DCC experiment involves several components, loosely aligned with the components of a system undergoing Darwinian evolution (Fig. 1.4): (1) a library of building blocks (components of a population), (2) a reversible reaction (analogous to a mutagenesis method or reproduction), (3) a selection mechanism, and (4) an analytical method. The relatively short history of DCC has seen many innovative approaches to

---

[1] Other terms have been employed for this general concept, including "self-assembled combinatorial libraries," "constitutional dynamic chemistry," and "virtual combinatorial libraries". "Dynamic combinatorial chemistry" and "dynamic combinatorial library" seem to have found the broadest usage, while "virtual combinatorial library" is perhaps best reserved for conditions under which library members form at concentrations below detection limits in the absence of target (e.g., Reference 81).

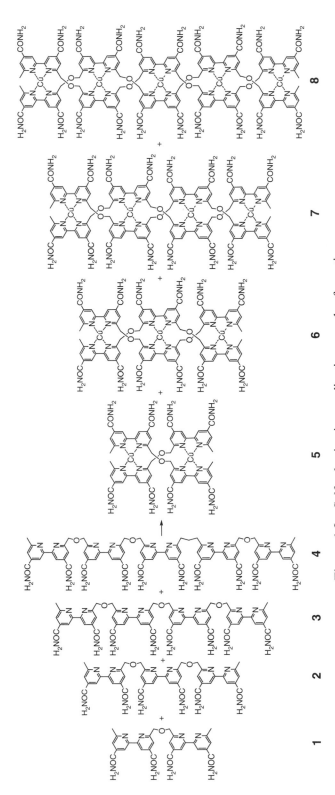

Figure 1.3 Self-selection in coordination complex formation.

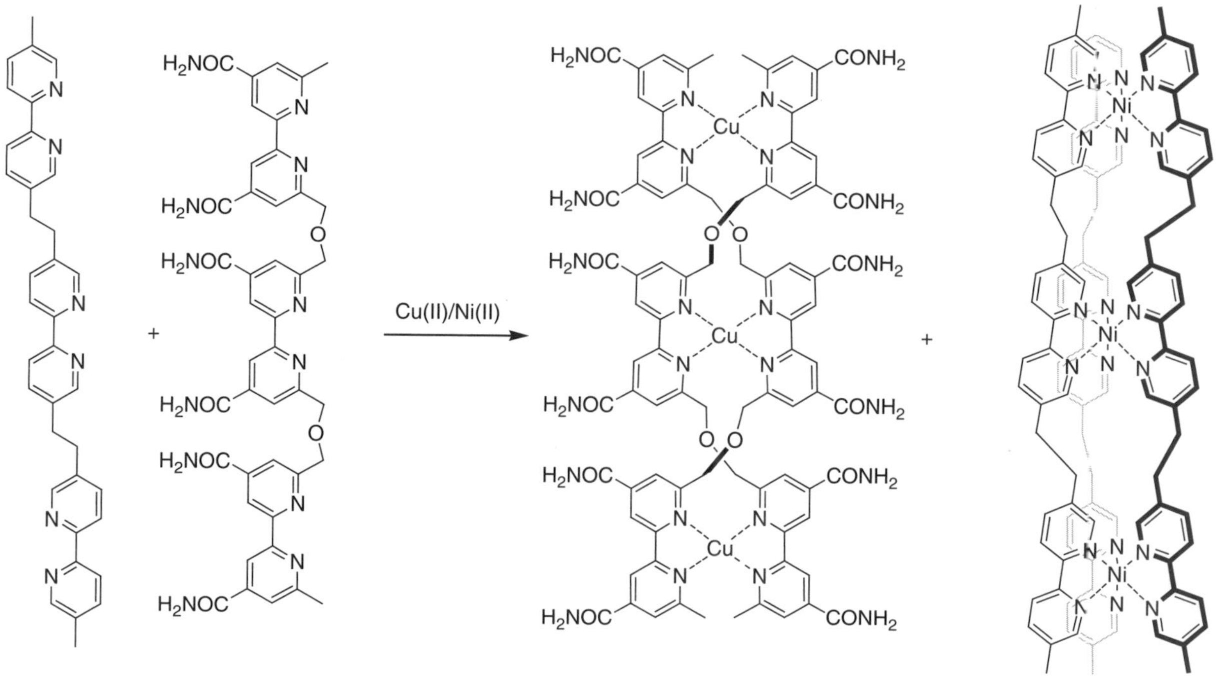

Figure 1.3 (Continued)

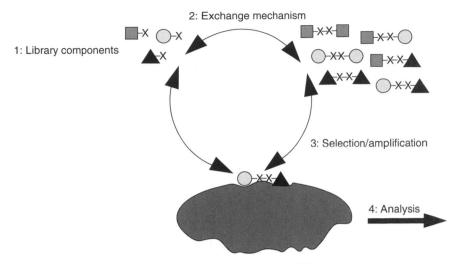

Figure 1.4   The basic structure of a DCC experiment.

each of these areas. They are all interrelated, making it somewhat difficult to discuss them in the linear fashion required by a chapter in a book. However, we will attempt to do so, by way of introduction to the field. This is not intended to be an exhaustive catalog of all dynamic combinatorial library (DCL) experiments, but rather an introduction to each topic via selected examples. More examples can, of course, be found in further chapters of this book.

1.2. Considerations in Choosing an Exchange Reaction

Chemists (and particularly synthetic organic chemists) have been trained to view synthetic reactions through one set of criteria: reactions should be irreversible and highly selective (but general). In contrast, DCC requires one to view candidate exchange reactions with a different set of criteria in mind. Most obviously, reactions must be reversible. Several other criteria are listed in the following text. Some of these constraints are particularly important when one is working with biomolecular targets.

1.2.1. Conditions Under Which Exchange Can Be Made to Occur

An obvious, but important criterion for selecting a scrambling reaction is that equilibration must occur under conditions compatible with the target.

For self-selection experiments and selection in the presence of an organic "guest," this is generally a simple criterion to satisfy. However, biomolecules dramatically narrow the available conditions: the reaction must ideally occur at room temperature and in buffered aqueous solution. Both of these conditions can be (at least in principle) attained by physically separating the scrambling reaction from the biomolecule against which the library is selected.

1.2.2. Rate of Exchange

The rate of exchange ideally needs to be fast enough that equilibrium is reached within a convenient interval, but slower than binding to the target. One certainly wants equilibrium to be reached faster than the target degrades, if biomolecular binding is the goal. In some cases, interesting things can occur in a very slow regime; for example, studies on folding-driven oligomerizaton by Moore and coworkers [13–15], in which imine metathesis was used as the exchange mechanism, required as much as 19 weeks or more to reach equilibrium [16], depending on the composition of the library.

1.2.3. Ability to Halt Equilibration

Once an equilibrium distribution of the library has been reached, one generally wants to be able to analyze this distribution in order to determine what compound has been amplified. This requires "freezing" the populations of individual library members such that the analytical method does not alter the composition of the library. Methods for halting (or at least greatly reducing the rate of) equilibration can include changes in temperature, changes in pH (disulfide exchange, imine metathesis, acetal formation), turning off of the light (*cis/trans* olefin isomerization), ablation of the reactive functionality (imine reduction), or removal of catalyst (olefin metathesis and other transition metal-catalyzed processes).

1.2.4. Selectivity

Talking about selectivity in the context of a combinatorial library seems odd, and indeed, from the perspective of generating maximum diversity, it is critical that the reaction is nonstereoselective (stereorandom) and non-substrate selective (general). However, it is important that reaction occurs only with desired functional groups on library constituents rather than with target functionality, or library functionality, leading to irreversible formation of a product.

Fulfilling all of these criteria is difficult, and to date only a very small subset of the reactions available for chemical synthesis has been employed in DCC experiments. In the following sections, we will discuss representative examples of exchange reactions that have proven successful; many others are described in other chapters of this book. Discovery of new types of exchange reactions remains one of the most important challenges in the field.

1.2.5. Disulfide Exchange

Disulfide exchange has proven to be one of the simplest, most robust, and most widely used methods for library equilibration. Extensive studies by the Whitesides group [17] and others in the late 1970s and early 1980s established that thiolate–disulfide exchange was facile in aqueous solution at slightly above neutral pH, but slow at neutral pH and below. The first use of disulfide exchange in a DCL, of which we are aware, was reported by Hioki and Still in 1998 [18]. Building on prior work in Still's laboratory on the design and synthesis of artificial receptors for peptides [19], the authors first examined the disproportionation of compound **13** in chloroform in the presence of 2 mol% thiophenol and triethylamine (Fig. 1.5). In the absence of target resin-bound peptide, equilibrium was reached at 35% **13**S–SPh and 65% PhS–SPh and **13**S–S**13**. However, after incubation with resin-bound Ac(D)Pro(L)Val(D)Val, the equilibrium shifted to 95% PhS–SPh and **13**S–S**13**, a change in K_{eq} from 3.8 to 360. Challenging the selection process with a somewhat more subtle mixture, Hioki and Still next examined the disproportionation of the mixed disulfide **13**S–S**14** in the presence of 10 mol% **14**SH and triethylamine. Although the shift in equilibrium composition was not quite as large in this case (evidence for some peptide-binding ability on the part of receptors including **14**SH in their makeup), it was still definitive: 75% **13**S–S**13** on the resin phase (bound to the peptide), and 85% **14**S–S**14** in solution.

Since this initial report, disulfide chemistry has become perhaps the most widely employed method of component exchange in DCLs. Disulfide exchange is rapid, and conducted under conditions ideal for library selection in the presence of biomolecules. It is highly suitable for even very complex libraries, as in the >11,000-compound resin-bound DCLs targeting RNA binding developed by the Miller group (described in detail in Chapter 3) [20,21], and in a >9000-compound solution-phase DCL reported by Ludlow and Otto [22], described in greater detail in the following text in the context of analytical methodology.

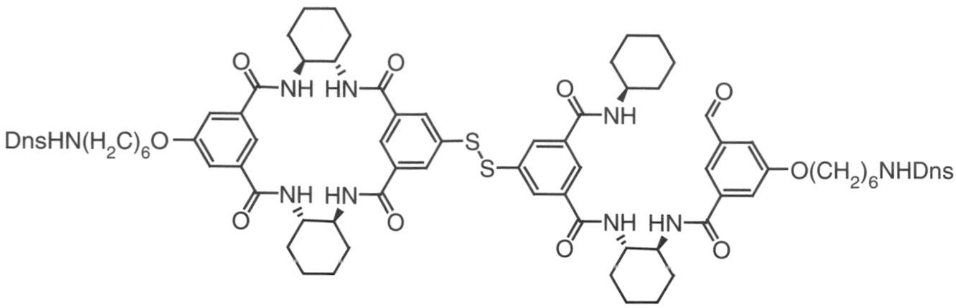

13   **13**

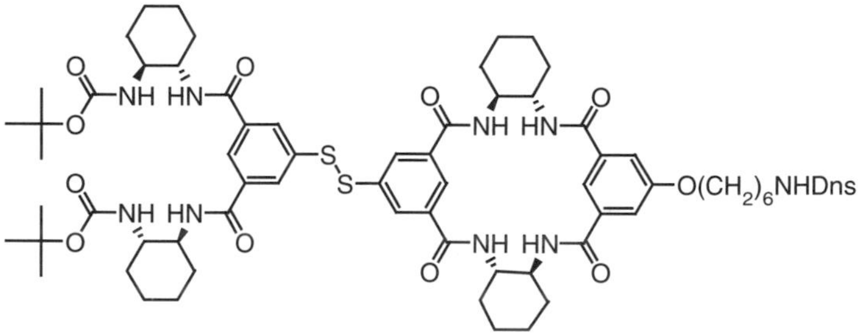

14   **13**

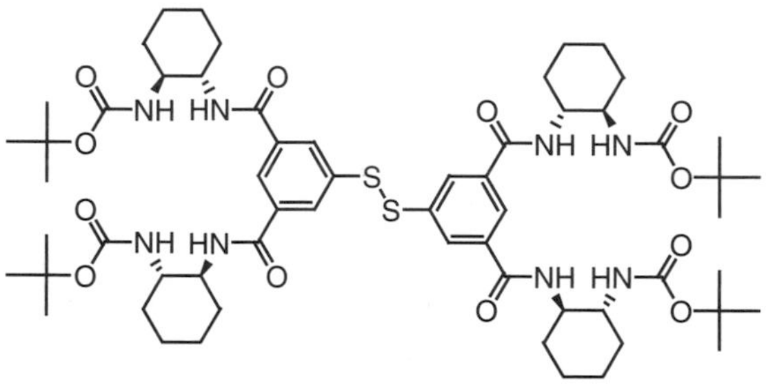

14   **14**

Figure 1.5 Disulfide-containing receptors for peptides prepared by Hioki and Still.

1.2.6. Imine Metathesis and Related Processes

As we have already mentioned, the ability of imine formation to serve as a useful reaction in templated systems was observed by Lynn et al. in the early 1990s. Use of imine metathesis in DCC was first described by Huc and Lehn in 1997 in a library targeting the production of carbonic anhydrase inhibitors [23]. In this case, reduction of the imines with sodium cyanoborohydride was employed to halt library equilibration. The authors noted that because of the 18 lysine ε-amino groups, in addition to the terminal amine, it was necessary to use an excess (15-fold) of starting amines in order to limit reaction between starting library aldehydes and the enzyme. Equilibration of the library in the presence of target carbonic anhydrase and NaBH$_3$CN was allowed to proceed for 14 days. HPLC analysis revealed strong amplification of compound **15**; this amplification did not occur in the presence of a competitive carbonic anhydrase inhibitor.

Imine metathesis has continued to be a popular exchange reaction for DCLs. Various groups have found novel systems in which the reaction can be applied, as well as interesting ways to halt the equilibration. For example, Wessjohann and coworkers have demonstrated that Ugi reactions can efficiently halt equilibration of an imine DCL, combining an irreversible diversification process with a reversible library selection [24]. Xu and Giuseppone have integrated reversible imine formation with a self-duplication process [25], and Ziach and Jurczak have examined the ability of ions to template the synthesis of complex azamacrocycles [26]. The mechanistically related reactions of hydrazone [27] and oxime [28] exchange have also been explored as suitable foundations for DCL experiments.

Another process mechanistically related to imine exchange is the dynamic production of pyrazolotriazinones reported in 2005 by Wipf and coworkers [29]. After first verifying that reaction of either **16** or **17** with equimolar quantities of isobutyraldehyde and hydrocinnamaldehyde at 40°C in water (pH 4.0) resulted in the same 3:7 mixture of **16** and **17** at equilibrium (Fig. 1.6, Eq. 1), the authors demonstrated that a library could be generated by reaction of pyrazolotriazinone **16** with a series of aldehydes (Fig. 1.6, Eq. 2). Direct metathesis of pyrazolotriazinones was also demonstrated, as was reaction with ketones. Importantly, equilibration was halted by raising the pH to 7.

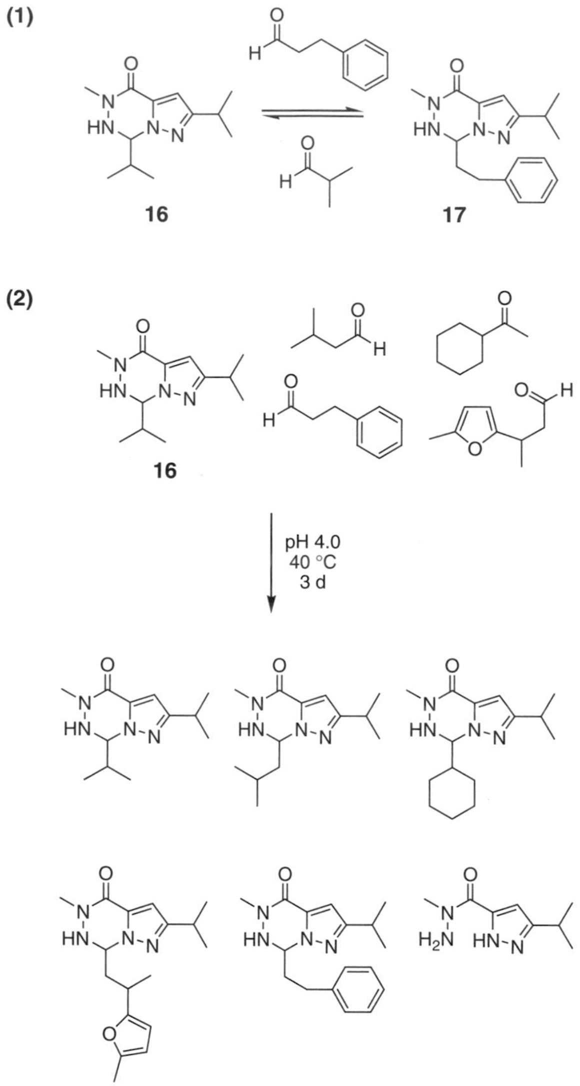

(1)

16   **17**

(2)

16

pH 4.0
40 °C
3 d

Figure 1.6   Pyrazolotriazinone metathesis (Wipf and coworkers).

1.2.7. Acetal Exchange

The reaction of aldehydes with alcohols to form acetals is rapid and reversible, and both the rate and the position of acetal–aldehyde equilibria can be affected by the pH of the reactant solution [30–32]. Thus far, however, relatively few studies have made use of transacetalization as

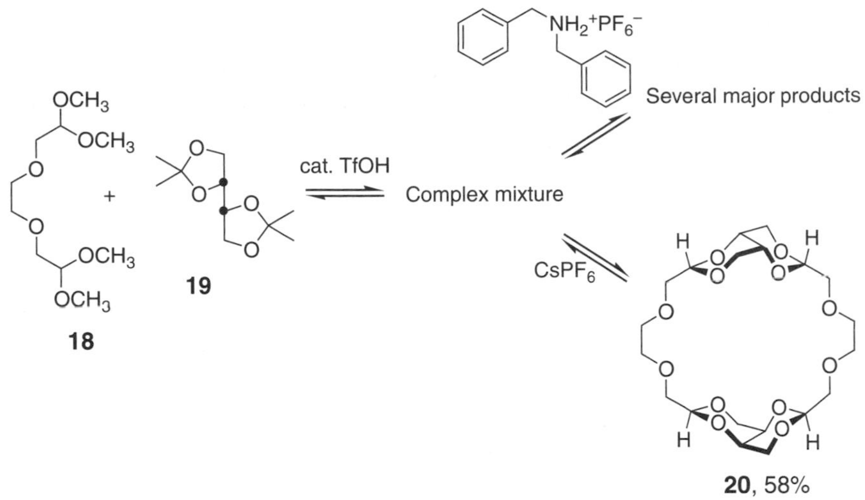

Figure 1.7 Transacetalization as a DCC exchange mechanism.

an exchange reaction in DCC. An initial demonstration of guest-induced equilibrium shifting in a library of acetals undergoing exchange was provided by Stoddart and coworkers in 2003 [33]. Treatment of a deuterochloroform solution of diacetal **18** and the D-threitol-derived acetonide **19** (rather than threitol directly because of threitol's low solubility in organic solvents) with catalytic TfOH initiated production of a library of cyclic and oligomeric acetals (Fig. 1.7). Addition of the hexafluorophosphate salt of dibenzylamine caused the population of species in the library to shift, attaining equilibrium after 3 days at 45°C. Although the authors reported a much simpler mixture, consisting primarily of [2+2] "macropolycycles" (i.e., cyclic structures derived from two molecules of **18** and two of **19**), NMR spectroscopy indicated that several isomers were present. In contrast, library selection conducted in the presence of $CsPF_6$ produced cyclic acetal **20** as the primary product, in 58% yield.

Dynamic transacetalization experiments targeting cyclophane formation have also been described by Mandolini and coworkers [34]. Production of a cyclic polyether DCL by direct reaction of triethylene glycol and 4-nitrobenzaldehyde has been reported by Berkovich-Berger and Lemcoff; amplification of small macrocyclic members of the library by ammonium ion was observed [35]. With these few examples demonstrating feasibility, we can anticipate increased use of transacetalization in future DCC efforts.

1.2.8. Transesterification

The Sanders group provided several early examples of thermodynamic self-selection from libraries, employing transesterification as the exchange reaction. In one example, the cholic acid methyl ester derivative **21** was induced to form an equilibrating mixture of linear and cyclic oligomers via refluxing in toluene in the presence of potassium methoxide–crown ether complex (Fig. 1.8) [36]. Equilibrium mixtures derived from cholic acid derivatives bearing R_2 = MEM, R_1 = OBn, or R_1 = R_2 = PMB strongly favored production of the cyclic trimer over that of other cyclic oligomers; R_1 = R_2 = H also yielded cyclic dimer. Related studies from the Sanders group likewise explored equilibrium selections derived from transesterification of cinchona alkaloids [37,38].

A closely related process is the equilibration of thioesters, explored by the Gellman group in the context of evaluating peptide stability [39]. Larsson and Ramström have also employed thioester exchange in the context of libraries targeting hydrolases [40], while Sanders, Otto, and colleagues have demonstrated that thioester exchange can operate in tandem with disulfide exchange [41]. Importantly, one can also decouple the thioester and disulfide exchange processes to allow for independent staging of the two.

1.2.9. Metal-Catalyzed Allylic Substition

Metal-catalyzed allylic substitution reactions have been a mainstay of synthetic chemistry because of their ability to proceed irreversibly and with high selectivity [42]. It is also feasible, however, to produce analogous systems that are completely reversible and nonselective, or ideally situated for use in DCC. These are essentially metal-catalyzed transesterification reactions, with the added feature of potentially providing stereochemical scrambling (and selection) as well as constitutional variation. An early example of this was provided in 2000 by Kaiser and Sanders [43]. In the absence of a template, reaction of diallyl diacetate **22** with a dicarboxylic acid in the presence of catalytic Pd(0) produced a negligible amount of the cyclized compound **23** (Fig. 1.9). However, when templated with 1,3-bis(4-pyridyl) benzene, yield of the cyclic structure increased to roughly 10%, independent of the dicarboxylic acid used.

In 2000 the Miller group provided a proof-of-principle study of Pd pi-allyl chemistry for library selection in the presence of a biomolecule [44]. In this approach, Pd(0) chemistry was employed to generate a library of cyclopentene-1,4-diesters in halogenated solvent (Fig. 1.10). This was allowed to equilibrate across a dialysis membrane with an enzyme target (pepsin) in buffered aqueous solution. LC-MS analysis of the library allowed identification of compound **24** as a library member amplified in the presence

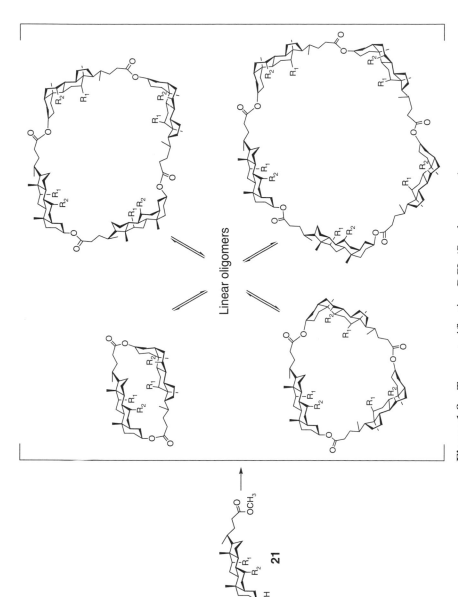

Figure 1.8 Transesterification DCL (Sanders group).

Linear oligomers

15

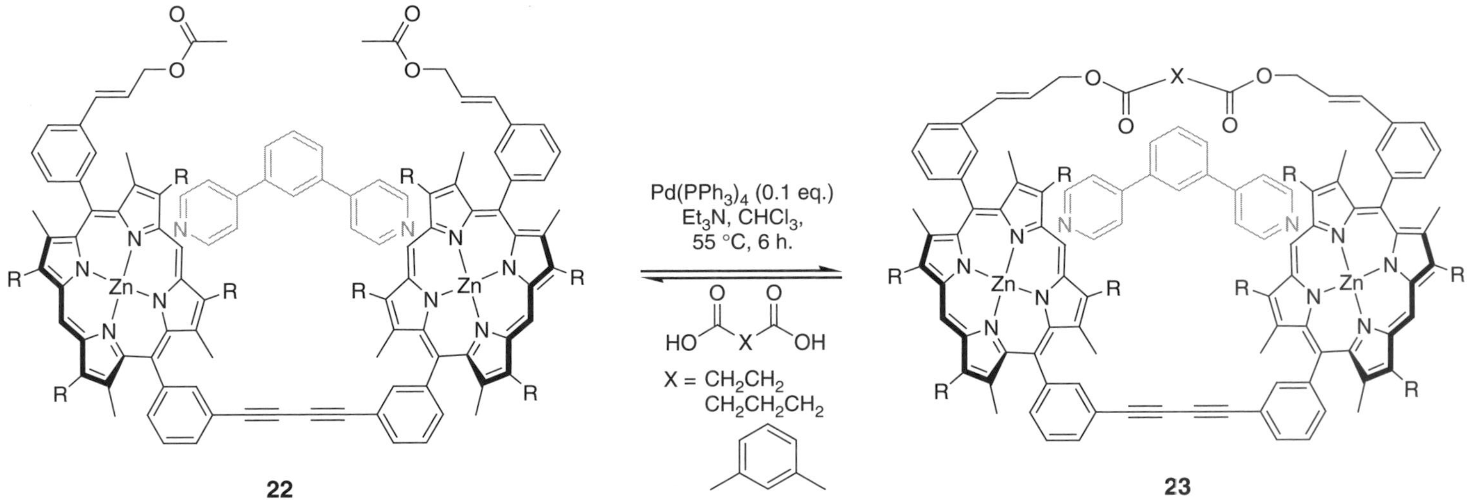

Figure 1.9 Pd pi-allyl-mediated self-selection (Sanders).

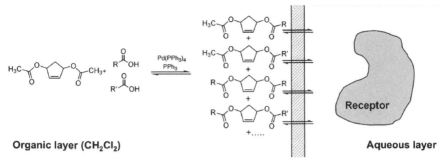

Figure 1.10 Biphasic pi-allyl Pd-based DCC selection for receptor binding.

24

of the enzyme. An obvious challenge in these systems is the role played by log P in the library selection process. Future studies in this area would include tuning of the solubility of library building blocks to provide sufficient solubility in chloroform for rapid exchange chemistry, while retaining the ability to remain in aqueous solution and bind to the target receptor. A recent demonstration of Pd-mediated pi-allyl sulfonylation in water [45] suggests that future "water-only" selection experiments may be possible.

1.2.10. Olefin Metathesis

Olefin metathesis is notable as one of the few exchange reactions of carbon–carbon bonds employed to date (the Diels–Alder reaction is the other primary example; *vide infra*). An early use of the metathesis reaction for capturing an equilibrating mixture of self-assembled structures was provided by the Ghadiri group. The cyclic peptide cyclo[–(L-Phe-D-(CH₃)NAla-L-Hag-D-(CH₃)NAla)2–] (**25**, where L-Hag is L-homoallylglycine) was designed to allow for interconversion between diastereomeric hydrogen-bonded dimers **26a** and **26b** (Fig. 1.11). As anticipated by the authors, treatment of a chloroform solution of **25** with Grubbs' first-generation catalyst provided cyclized structure **27** in 65% yield.

The Ghadiri work set the stage for later experiments employing olefin metathesis in a library selection.[2] The Nicolaou group reported the first of

---

[2] As well as similar demonstrations of covalent capture of hydrogen-bonded dimers (e.g., see Reference 46).

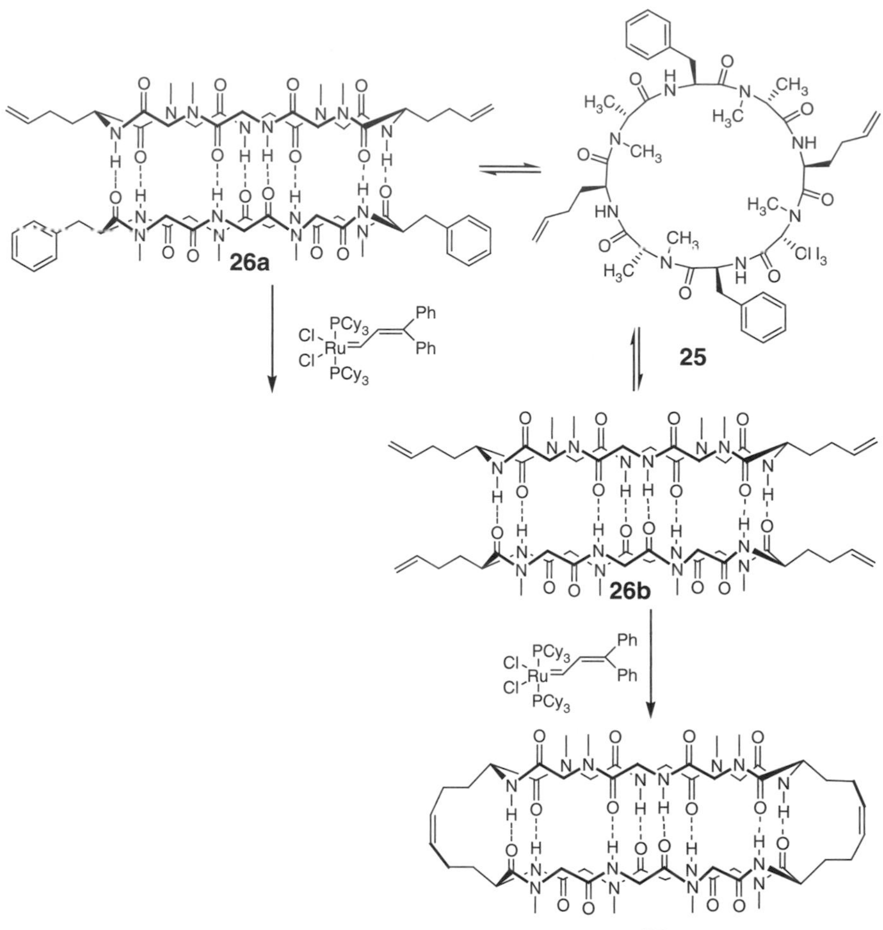

Figure 1.11   Covalent capture of peptide macrocycle dimer (Ghadiri).

these in 2000, in a study focused on the production of vancomycin dimers. A particularly interesting aspect of this work was that both olefin metathesis and disulfide exchange were examined, thus allowing a comparison of the results with the two. Although not intended as DCL experiments per se (the work was described as "target-accelerated synthesis", and no attempt was made to examine whether libraries reached equilibrium), the experiments nonetheless provided an interesting demonstration of the potential DCC application of cross-metathesis. Initial experiments established the ability of either Ac-D-Ala-D-Ala or Ac_2-L-Lys-D-Ala-D-Ala to accelerate the production of vancomycin dimer **29** via cross-metathesis of **28** (Fig. 1.12); subsequent library experiments allowed optimization of the tether linking

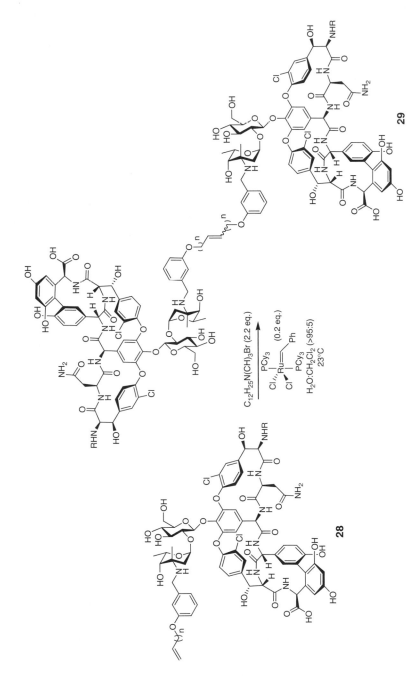

Figure 1.12 Target-accelerated synthesis of vancomycin dimers via olefin metathesis (n = 2 or 4).

the two halves of the dimers. Importantly, amplification (or acceleration) of dimers based on their affinity for the peptide targets correlated well with bactericidal activity. A recent DCC experiment targeting vancomycin analogs has been reported employing resin-immobilized peptides (by analogy to the work of Hioki and Still) by Chen et al. [47].

Like all reversible reactions, systems employing olefin metathesis are subject to self-selection. This was examined in the context of self-metathesis of simple allyl- and homoallylamides by McNaughton et al. [48]. In that report, both the yield of self-metathesis products and the ratio of *cis-* and *trans-*olefin isomers produced depended strongly on remote functionality on the homoallylamide. A 2005 study by Nolte, Rowan, and coworkers [49], focused on the templated production of porphyrin "boxes," provides an interesting case study in the need to also carefully consider issues of catalyst reactivity in the design of metathesis-based DCC experiments. As shown in Fig. 1.13, the authors first subjected zinc porphyrin derivative **30** to Grubbs' second-generation metathesis catalyst in the presence of tetrapyridylporphyrin template **31** (TPyP). Contrary to expectation, this provided only a low yield of the desired TPyP-coordinated cyclic tetramer, instead of providing a complex mixture of products. The low yield was attributed to coordination of TPyP to the ruthenium catalyst. In contrast, treatment of **30** with Grubbs' first-generation catalyst to produce a library of cyclic and linear oligomers, followed by re-equilibration of the library in the presence of the TPyP template, yielded the desired structure in substantially higher yield.

1.2.11. Alkyne Metathesis

Alkyne metathesis, a mechanistic cousin of alkene metathesis, has thus far found only limited exploration. In 2004, Zhang and Moore reported that precipitation-driven alkyne metathesis reactions could efficiently produce arylene ethynylene macrocycles [50]. This was explored further in a 2005 report, verifying that the products obtained were indeed the result of thermodynamic self-selection [51].

1.2.12. Diels–Alder Reaction

Joining olefin metathesis on the very short list of exchange reactions involving carbon–carbon bonds, the Diels–Alder reaction was studied in 2005 by Lehn and colleagues [52]. As the authors note, most Diels–Alder reactions proceed only in the forward direction at room temperature, with retro Diels–Alder reactions typically requiring elevated temperatures. Careful tuning of the diene and dienophile, however, can alter this significantly. In particular, reactions of substituted fulvenes (**32**) with diethylcyanofumarate (**33**) were found to reach an equilibrium mixture of cycloadduct products and starting materials "within seconds" of mixing at room temperature in chloroform (Fig. 1.14). Reversibility of the reaction was established through a series of diene exchange reactions.

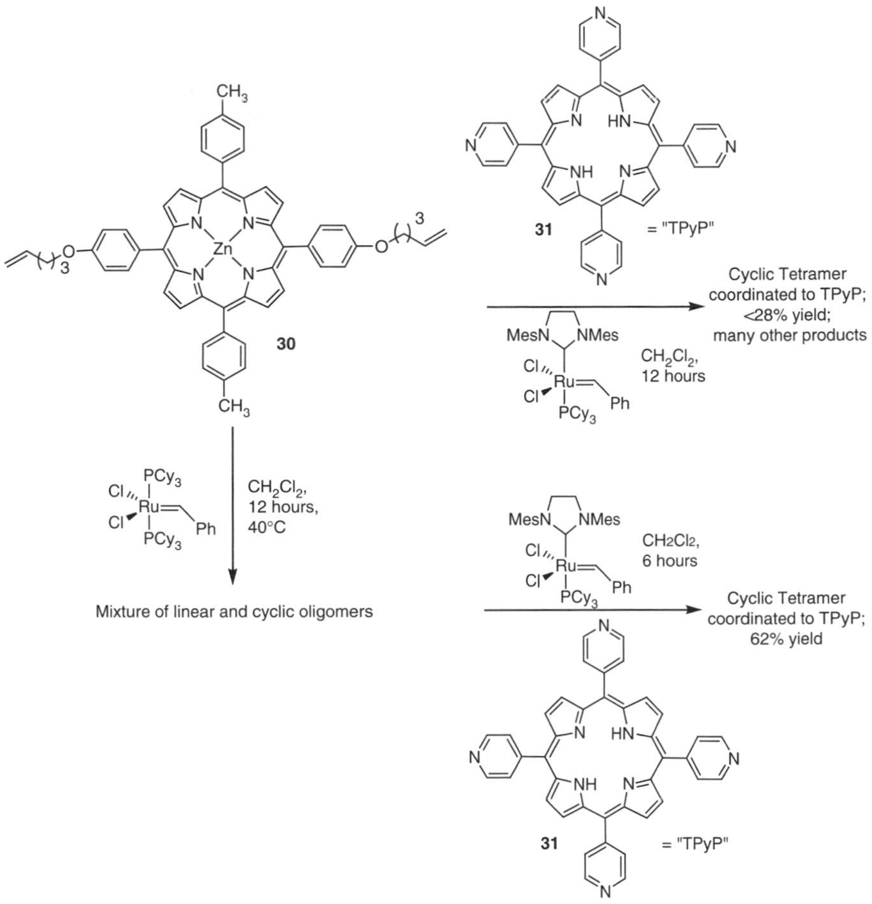

Figure 1.13   Self-selection of molecular boxes via olefin metathesis.

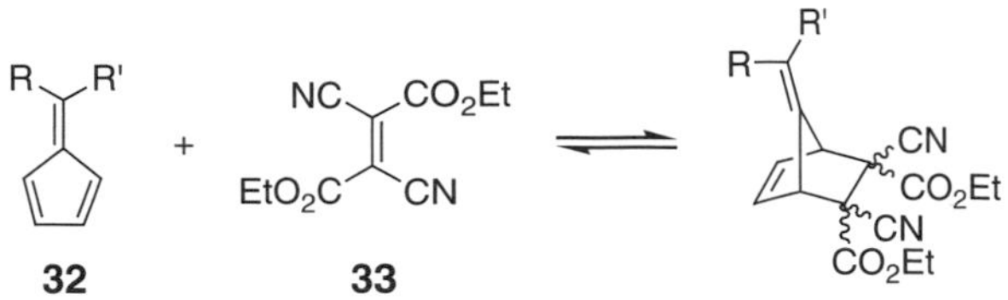

Figure 1.14   Reversible Diels–Alder reaction of substituted fulvenes with diethylcyanofumarate.

Bennes and Philp have employed a simple DCL based on a reversible Diels–Alder reaction to study kinetic versus thermodynamic selectivities, as well as concentration-dependent compensatory effects in a DCL self-selection process [53]. Rate constants and equilibrium constants for the reaction of dienophiles **34**, **35**, and **36** (Fig. 1.15) with diene **37** in $CDCl_3$ were first established. These confirmed molecular modeling predictions that cycloaddition between **37** and the dienophile **34** bearing a two-carbon spacer provided the most thermodynamically stable product, presumably because of an ionic or a hydrogen-bonding interaction between the carboxylic acid and amidopyridine moieties. Interestingly, reaction between **37** and the one-carbon spacer dienophile **34** had the highest rate, however (kinetically favored product). Running the reaction as a DCL provided the counterintuitive observation that maximum selectivity for the thermodynamic product **38** was obtained at low conversion. This is hypothesized to result from a compensatory effect: as dienophile **35** is depleted from the pool of available reactants, more of the less thermodynamically stable products are formed simply because of differences in the concentration of reactant dienophiles. This effect has also been studied extensively by the Severin and Otto groups (among others) and is discussed further in the following sections.

1.2.13. Photochemical Isomerization

While the vast majority of DCC experiments have focused on equilibration of constitutionally distinct library members, methods for the

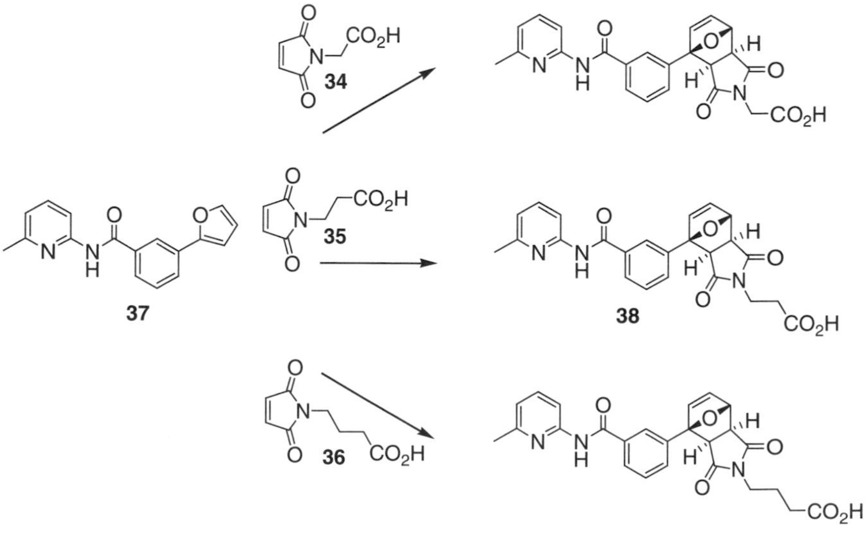

Figure 1.15   Recognition-mediated selectivity in a reversible Diels–Alder DCL.

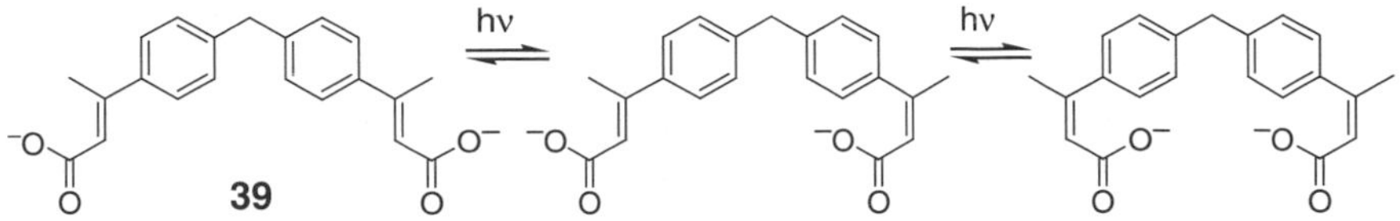

Figure 1.16   Photochemical isomerization (Eliseev).

equilibration of configurational isomers are also of interest. An early example of DCC from the Eliseev group employed photochemical *cis/ trans* isomerization as the exchange reaction [54]. As shown in Fig. 1.16, photolysis of dicarboxylate **39** yields a mixture of three isomers, with a photostationary state of 17:31:52 *trans/trans*:*cis/trans*:*cis/cis*. Subjecting this mixture to 30 cycles of irradiation followed by passage through an affinity column bearing guanidinium groups (as the selection process), and subsequent elution of material on the column, yielded a substantially altered mixture: 2:13:85 *trans/trans*:*cis/trans*:*cis/cis*.

Surprisingly, some 11 years would elapse before another example of the use of photochemical *cis/trans* isomerization as a diversity-generating reaction in DCC would appear in the literature. In a 2008 report [55], Ingerman and Waters described the use of azobenzene photoisomerization and hydrazone exchange as a "doubly dynamic" system (further examples of multiexchange systems are presented below). Unlike the Eliseev work, photochemical equilibration was carried out in the presence of the target. As the authors note, photoequilibration converts the library to a photostationary state rather than a thermodynamic minimum, but binding to a particular library member can alter the distribution of products in the photostationary state just as readily as binding can alter the distribution of a thermodynamic equilibrium.

1.2.14. Metal Coordination

The ability of metal coordination to influence the distribution of materials formed in a labile mixture was recognized as early as 1927, in a pair of studies examining the self-condensation of aminobenzaldehydes [56,57]. As discussed above, many other experimental antecedents of DCC centered on observation of self-selection processes occurring during the formation of coordination complexes, and it is therefore not surprising that transition metal complexes capable of facile ligand exchange have been the subject of library experiments. A particular challenge in this case is that one must choose the coordination carefully, as many coordination complexes are too labile to permit simple analysis postequilibration. Indeed, some early

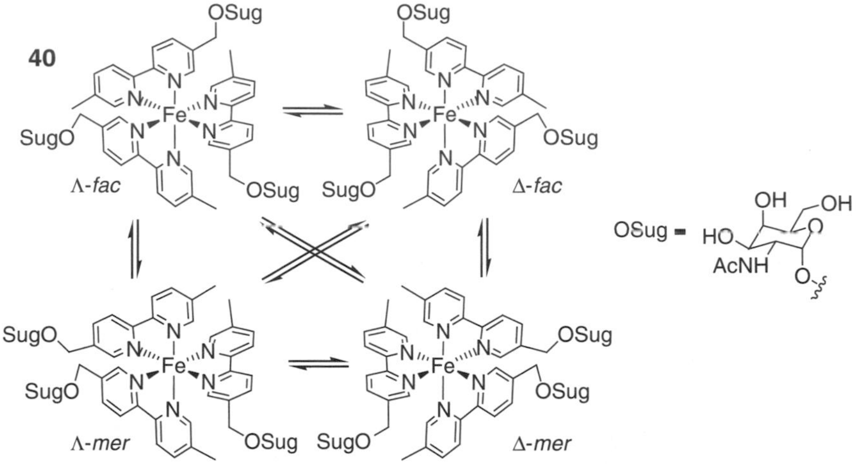

Figure 1.17   Stereomutation of Fe(II) complexes.

experiments from our group involved complexes whose existence could only be inferred based on analysis of stable derivatives [58,59]. However, more "cooperative" systems have been reported by others. For example, a 1997 report from Sakai, Shigemasa, and Sasaki explored the lectin-mediated selection of carbohydrate-based ligands from an equilibrating mixture of Fe(II) complexes [60]. In the presence of Fe(II), bipyridyl carbohydrate derivative **40** forms an equilibrating mixture of steroisomeric complexes, as shown in Fig. 1.17. Introduction of *Vicia villosa* B4 lectin causes this equilibrium to shift in favor of the Λ-*mer* isomer (from 29% of the mixture to 85%), which is best able to bind to the carbohydrate-binding site. In this case, individual complexes were sufficiently stable to permit analysis by HPLC.

Buryak and Severin have described the use of dynamic libraries of Cu(II) and Ni(II) complexes as sensors for tripeptides [61]. A notable aspect of this work is that as isolation of the metal complexes is not necessary (sensing is accomplished by observing changes in the UV-vis spectrum), potential concerns over the lability of coordination complexes do not apply. Specifically, three common dyes [Arsenazo I (**41**), Methyl Calcein Blue (**42**), and Glycine Cresol Red (**43**), Fig. 1.18] were mixed with varying ratios and total concentrations of Cu(II) and Ni(II) salts in a 4×5 array. Previous work had demonstrated that these conditions produced equilibrating mixtures of 1:1 and 2:1 homo- and heteroleptic complexes [62]. These arrays were able to clearly and unambiguously differentiate tripeptides based on the differential pattern of response. The Severin laboratory has

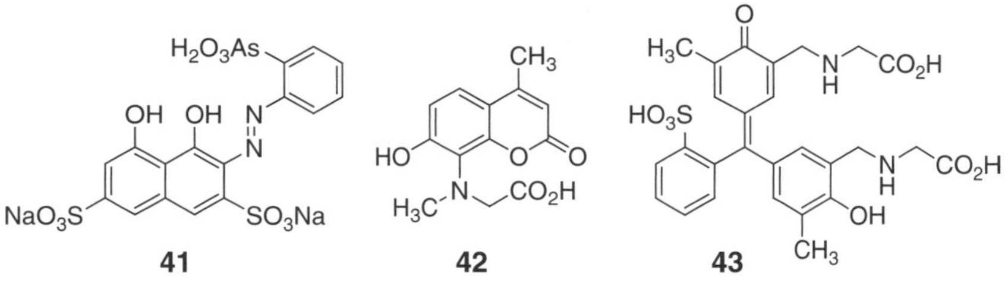

Figure 1.18 Dyes employed in the construction of Ni(II)/Cu(II) coordination DCLs for tripeptide sensing (Buryak and Severin).

successfully employed a similar strategy (which they call a "Multicomponent Indicator Displacement Assay", or MIDA) for nucleotide sensing [63] and as molecular timers [64].

Complex cage structures produced by the reversible assembly of pyridine 2-carboxyaldehyde, biphenyl amines, and iron salts have been described by the Nitschke group [65]. Interestingly, these were found to be capable of capturing hydrophobic solvent molecules as guests and carrying them into aqueous solution. Addition of a competing amine set off an imine exchange reaction that "unlocked" the cage complex, liberating the guest solvent.

1.2.15. Enzyme-Mediated Processes

Enzymes can also be brought to bear as catalysts for effecting scrambling reactions. This can be particularly useful in cases where the bond breaking/making process of interest is one not generally viewed as labile under conditions amenable to standard DCL experiments. For example, an early demonstration of DCC was provided by Swann et al., who employed thermolysin, a bacterial metalloprotease, as transamidation catalyst. Mixing H_2N-Tyr-Gly-Gly-COOH and H_2N-Phe-Leu-COOH with thermolysin resulted in the production of H_2N-Tyr-Gly-Gly-Phe-Leu-COOH, as well as other unidentified peptides. Incubation of this system with a target (fibrinogen, separated from the thermolysin solution by a dialysis membrane) amplified a fibrinogen-binding peptide relative to the rest of the mixture.

Enzyme-mediated chemistry can also inspire the development of novel nonenzymatic catalysts. Stahl, Gellman and collaborators at Wisconsin have taken on the challenge of developing transamidation catalysts, successfully identifying Al(III) complexes capable of equilibrating mixtures of tertiary carboxamides with secondary amines [66,67]. For example, treatment of an equimolar mixture of **44** and **45** with 2.5 mol% of an aluminum catalyst in

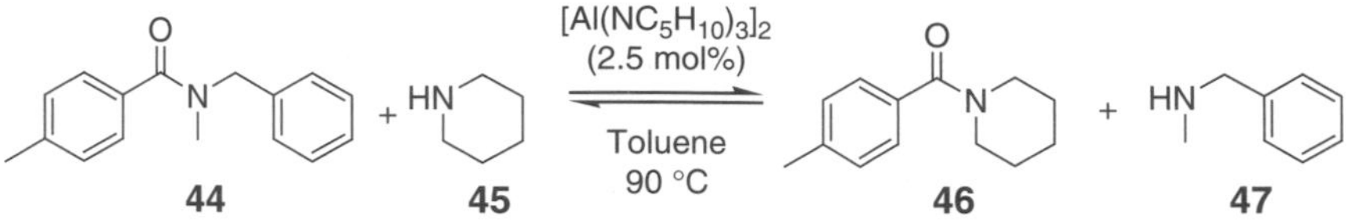

Figure 1.19   Aluminum-mediated transamidation in toluene.

toluene at 90°C rapidly affords a thermodynamic mixture of transamidated species **46** and **47** (Fig. 1.19). The exchange rate is first order in catalyst concentration, and independent of the concentrations of amine and amide. While the conditions employed are obviously incompatible with biomolecule-directed DCC, this nonetheless represents an important step forward and sets the stage for the development of catalysts capable of functioning closer to room temperature.

1.2.16. Multiple Exchange Reactions

One can in principle combine different exchange reactions in the same system in order to further increase the structural diversity accessed by the library. However, as this compounds the problem of selectivity (i.e., one now has two or more reactions that must exclusively involve one pair of functional groups), there are very few examples thus far of the practical implementation of this concept. An early, highly intriguing example was described by Lehn and coworkers in 2001 [68]. In this system, imine exchange (acyl hydrazone formation) and reversible metal coordination were employed in library generation.

The ability of boronic acids to serve as components of DCLs has been recognized for some time. For example, both the Shinkai [69] and Shimizu [70] groups have explored the properties of reversibly formed, oligomeric structures produced by reaction of bifunctional boronic acids with diols. In a recent example, the Severin group has demonstrated assembly of macrocycles via imine formation combined with the reversible reaction of boronic acids with diols (Fig. 1.20) [71]. Reaction of 3-formylphenylboronic acid, 1,4-diaminobenzene, and pentaerythritol provided macrocycle **48** as the primary characterizable product in 44% yield. Increasing the complexity of the system by addition of tris(2-aminoethyl)amine (tren) unfortunately produced a material of insufficient solubility to permit characterization, but changing the boronic acid from 3-formyl to the 4-formyl isomer allowed isolation and characterization of the cryptand **49**. Pushing the complexity of the self-selection process still further, Severin and coworkers mixed 3-aminophenylboronic acid, pentaerythritol, 3-chloro-4-formyl pyridine, and

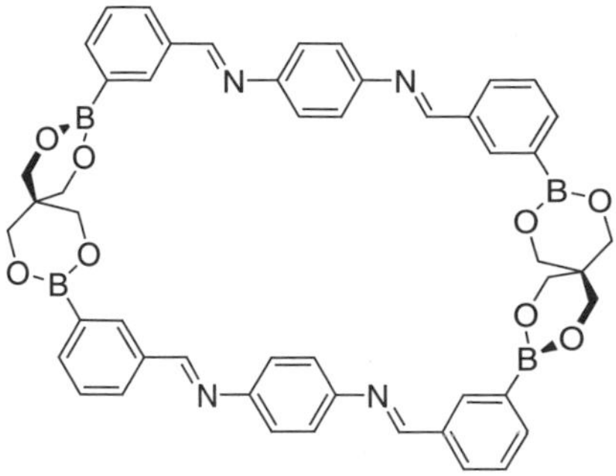

Figure 1.20 Multicomponent exchange.

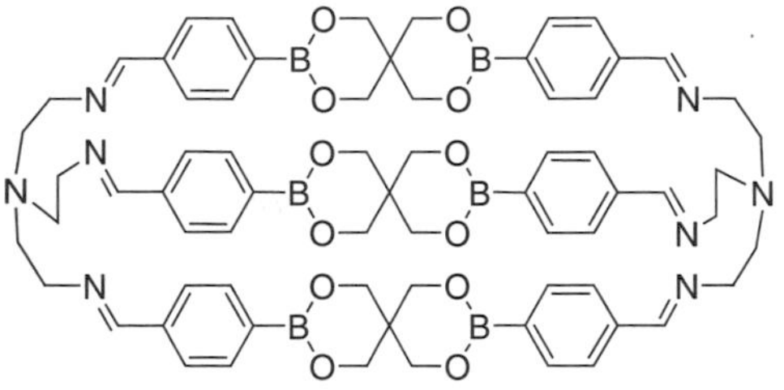

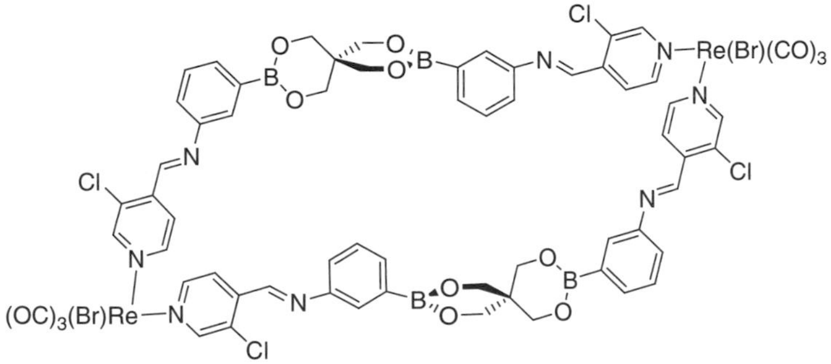

ReBr(CO)$_5$ to produce macrocycle **50**. This serves as an elegant proof-of-concept for incorporation of three-way orthogonal exchange reactions in DCLs. Several obvious challenges remain, however, as the conditions under which the reactions occur (refluxing THF/benzene) place obvious constraints on the targets that can be employed.

1.3. Library of Building Blocks

Once an exchange reaction has been chosen, the researcher must next choose a set of building blocks for construction of the dynamic library.

Considerations for building blocks in DCC experiments include the following:

Molecular weight: This is a two-part criterion, as both the absolute molecular weight and its uniqueness are important for each building block. The first of these is particularly important in the context of libraries focused on "drug-like" molecules, since one generally wants to keep the total molecular weight low. Uniqueness is critical if one plans to employ mass spectrometry for analysis of library results.

Solubility: Anecdotal evidence from several sources suggests that this is a particular concern. However, it is difficult to predict a priori, particularly in instances where oligomer libraries (rather than simple binary or A/B type libraries) are to be generated.

Structural diversity: One wants to be able to generate the greatest structural diversity possible with the smallest number of components. This both increases the chance of success in a binding-directed experiment (as opposed to a self-selection) and simplifies the analytical challenge.

Unique functionality: This is required for participating in the reversible reaction, and is the converse of the criterion listed above for choosing an exchange reaction. This functionality is carefully chosen to allow production of a binary A/B library, or formation of oligomers or cyclic structures. If multiple exchange reactions are anticipated, this increases the complexity of the design process accordingly.

Many of the issues one needs to consider in the selection of building blocks for DCC are common to all library experiments (including "static" libraries as well as DCC). One generally wants to generate as much structural diversity as possible from the smallest possible number of building blocks; the more diversity among the building blocks, the better the DCL is. If one is using DCC to target molecules for *in vivo* use, either as drugs or as probes, molecular weight can be an important consideration, since bringing DCL fragments together to form dimers (or trimers or oligomers) can obviously escalate molecular weight well beyond the size typically considered drug-like [72]. Molecular weight *uniqueness* is also an important consideration if mass spectrometry is intended as the primary analytical method for the library (discussed further in Chapter 7). Although unique molecular weights of DCL building blocks do not guarantee unique molecular weights for each member of the DCL, they clearly reduce the number of overlapping masses in the final library.

1.4. Selection Mechanisms

We have already discussed many different selection mechanisms in our brief survey of exchange reactions above. These can include ligand–receptor binding, self-selection, physical properties (of a polymer, etc.), and phase selection (binding to a target on solid phase, crystal packing). "Ligand–receptor binding" can be broken down into a number of smaller categories including DNA–small molecule, protein–small molecule, small molecule–small molecule ("host"–"guest"), ion–receptor, and others. It is also possible to combine the selection mechanism with the scrambling reaction in a negative selection, for example, by employing an enzyme capable of selectively destroying some library components. Such methodology is discussed further in Chapters 2 and 6.

1.5. Analytical Methodology

The problem of efficiently identifying "active" compounds in mixtures was noted early in the history of "static" combinatorial chemistry, and led to the development of many elegant strategies including binary encoding and recursive deconvolution. DCC, because of its integrated amplification step, should be less susceptible to the mixture problem. However, in practice, identification of the selected component can still present a challenge, particularly in cases where the theoretical size of the library is large. Typically, analytical strategies fall into one of two broad categories: (1) those independent of the selection scheme and (2) those coupled to the selection scheme.

Selection-independent analysis: In this case, library analysis occurs strictly after and apart from the library selection experiment. Typically, what this means is that the solution resulting from a library is analyzed by HPLC or HPLC-mass spectrometry (HPLC-MS), and compared with the chromatographic trace obtained for an identical library prepared in the absence of target. This provides an internal control for self-selection processes and (hopefully) allows direct identification library members undergoing enhancement through visual inspection. If self-selection is the goal, one simply compares HPLC traces of libraries at different time points.

The challenges of this method have kept the majority of DCLs relatively small. However, Ludlow and Otto recently demonstrated that, in some cases at least, direct HPLC-MS analysis of large libraries is possible [22].

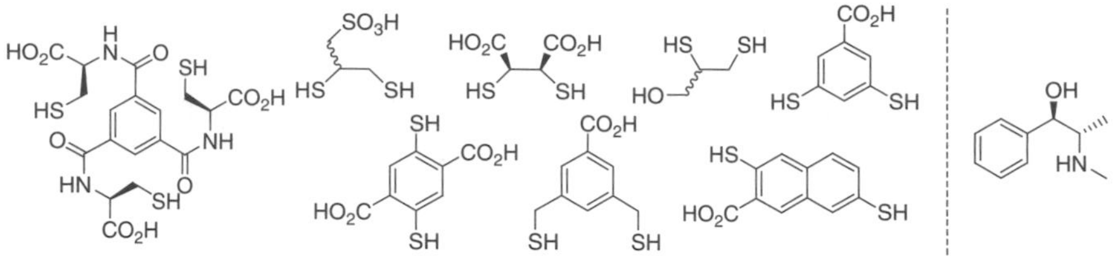

Figure 1.21   Components of a 9000-compound solution-phase DCL.

As shown in Fig. 1.21, a series of di- and tri-thiols were mixed under conditions suitable for disulfide formation and exchange, and allowed to evolve in the presence of an ephedrine template. HPLC-MS analysis of the library mixture after equilibrium had been reached allowed the identification of two heterotetrameric receptors with high ($K = 10^4$) affinity for ephedrine in borate buffer, although it is not clear whether these were in fact the "best" binders in the library.

Selection-coupled analysis/phase segregation: One strategy for simplifying the analytical challenge is to use phase segregation. Three subclasses are possible. In the first of these, a phase transition is part of the selection process. This includes not only the familiar crystallization-induced enantiomeric enrichment discussed above but also the experiments (primarily employed in experiments directed toward the production of novel materials) such as those described by Lehn and coworkers in 2005. In this study, an acylhydrazone library was created from guanosine hydrazide and a mixture of aldehydes (Fig. 1.22); in the presence of metal ions, formation of G-quartet structures led to the production of a gel.

Liquid–liquid phase segregation has been accomplished using two immiscible solvents (i.e., "phase transfer" DCC) by several laboratories. For example, the Morrow group has reported on imine [73] and acylhydrazone [74] DCLs targeting extraction of metal ions from aqueous to halogenated solutions. As discussed above in the context of Pd-mediated transesterification, the Miller group has also contributed to this area.

An alternative formulation of the phase-transfer DCC concept was reported in 2008 by the Sanders group [75]. In this case, thiol monomers were dissolved in water on either side of a U-tube containing chloroform (Fig. 1.23). After allowing the system to reach equilibrium, monomer distribution was identical in both aqueous solutions, and mixed species (e.g., **51**) were observed in the chloroform layer.

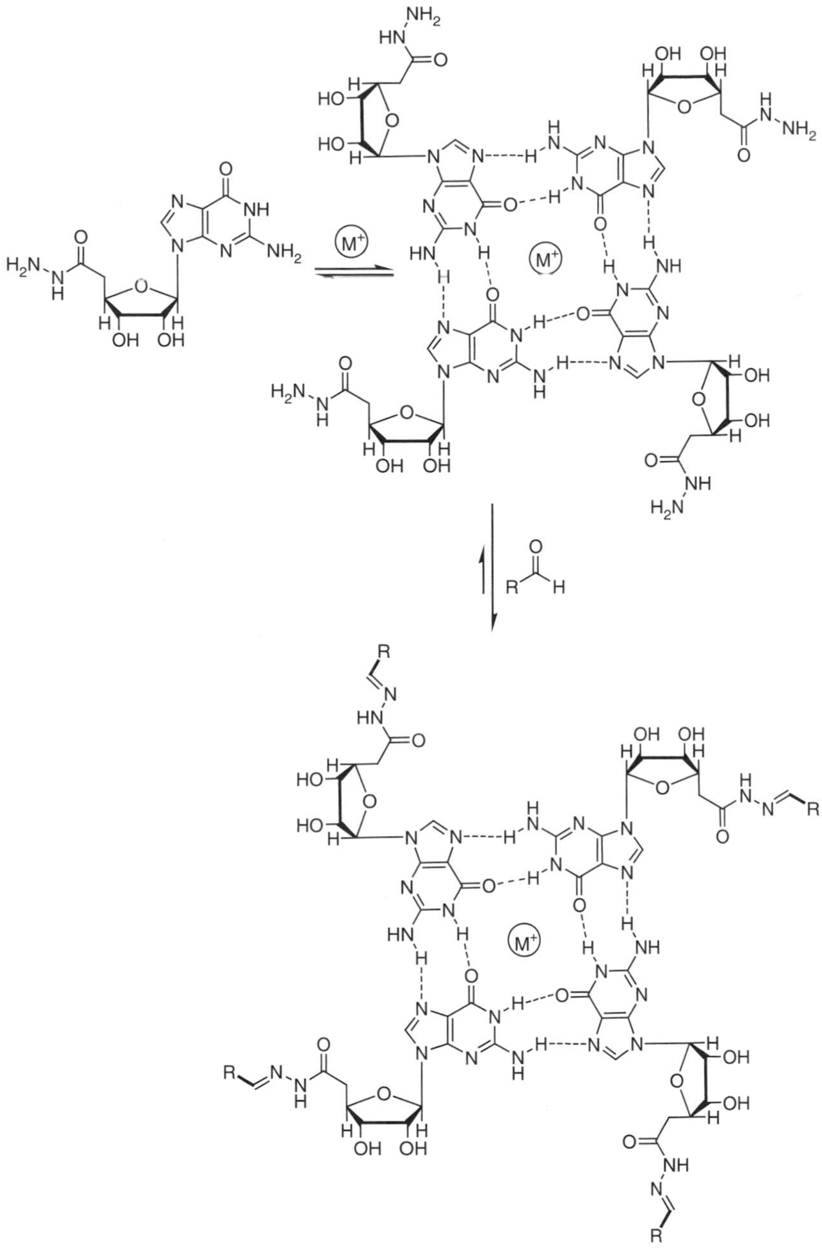

Figure 1.22   Selection by gelation (Lehn and coworkers).

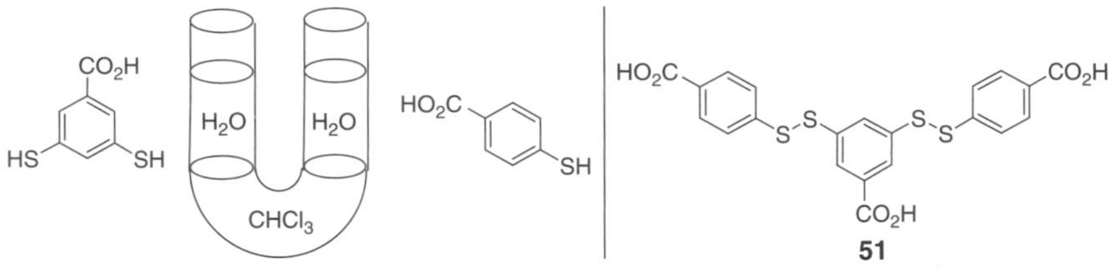

Figure 1.23   Phase-transfer DCC (Sanders).

A second method of incorporating phase into the selection process is to immobilize the target. Essentially an affinity chromatographic method, this allows nonbound library constituents to be washed away, leaving the selected compound(s) bound to resin. Eliseev's guanidine resin and Still's peptide-bearing beads, both discussed above in the context of exchange reactions, are examples of these.

Finally, one can also invert the affinity chromatographic concept and immobilize the library constituents themselves. We have termed this technique "resin-bound DCC" (RBDCC), and have found it to be an exceptionally efficient method for generating and screening large DCLs. The RBDCC method evolved out of an earlier attempt to phase-tag library components, in this case using a microarray format. This is illustrated schematically in Fig. 1.24, using a dimer library A_n-B_m as an example. One would first create an array of all A_n and B_m on the chip (in this case, the y-axis of the array would just be replicate spots of the individual library components; one could also imagine printing a single row). After carrying out a control experiment to verify that individual array-immobilized monomers did not bind to the fluorophore-labeled target, one would introduce an identical set of A_n and B_m in solution and allow the library to evolve in the presence of the target. Once equilibrium was reached, one could then wash the array and directly identify the selected components based on a simple imaging experiment. For example, if only the row corresponding to A_1 showed visible fluorescence, then one would conclude that only A_1-A_1 was the active compound. However, if both the A_1 and B_3 columns fluoresced, one would then have to evaluate three possible binders: A_1-A_1, A_1-B_3, and B_3-B_3. Of course, a major assumption of this method is that immobilization on the chip does not diminish the ability of any library component to participate in binding.

In practice, this array-based method is ineffective, primarily because there is insufficient material on the surface of the array to compete with solution-phase library members. As discussed in Chapter 3, however, implementing the RBDCC concept on resin beads produced a viable method for

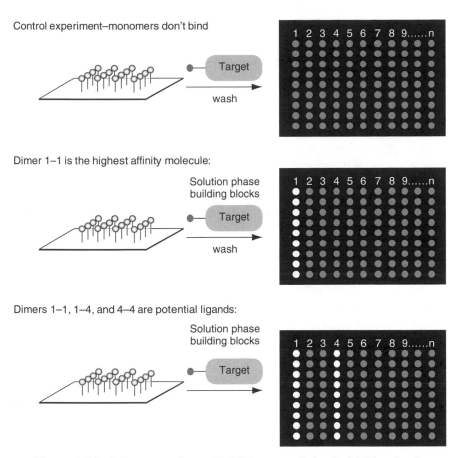

Control experiment–monomers don't bind

Target

wash

1 2 3 4 5 6 7 8 9......n

Dimer 1–1 is the highest affinity molecule:

Solution phase
building blocks

Target

wash

1 2 3 4 5 6 7 8 9......n

Dimers 1–1, 1–4, and 4–4 are potential ligands:

Solution phase
building blocks

Target

1 2 3 4 5 6 7 8 9......n

Figure 1.24   Microarray-format DCC (courtesy Brian R. McNaughton).

the generation of large DCLs and simple identification of "best" binders to biological targets. It is still conceivable that one might make the array-based method work through the use of extremely low solution volumes, and this will be an interesting area of exploration for the future.

1.6. Simulation of DCL Behavior

Many of the early "proof of concept" DCC experiments were carried out on a somewhat ad hoc basis. Several authors quickly came to the realization that it would be useful to develop methodology for simulating DCLs, both as a guide to experimental design and as a way to resolve the question of whether screening a DCL really leads to the identification of the library method with optimum fitness for the selected property (i.e., tightest binder).

The earliest efforts toward simulating the behavior of DCLs and DCL selection processes were reported by Moore and Zimmerman [76]. In this relatively simple model, the authors focused on predicting the behavior of a large population (e.g., 10^{10}) of interconverting copolymer sequences. Using a compilation of binding constants originally developed by Connors for cyclodextrin complexes [77], binding for the theoretical system was modeled as a normal (Gaussian) distribution in log K. Given a standard deviation in the binding constant of one order of magnitude, the model suggested that the mean binding constant for the population (i.e., for the library as a whole) could be shifted by no more than two orders of magnitude by a typical DCC selection process. This in turn led to the overall conclusion that although DCC could be a useful method for identifying lead compounds, it would not provide a practical method of synthesizing ultra-high affinity molecules unless the selection process could be coupled to an amplification method more powerful than equilibrium shifting. Of course, as the largest DCL synthesized to date is on the order of 10^4 (rather than 10^{10}), it is not clear how precisely this model will correlate with common DCC experiments.

While Moore and Zimmerman considered the behavior of a DCL as a whole, more recent efforts by the Otto, Sanders, and Severin laboratories have attempted to model the concentrations of each library component explicitly.[3] Notably, these groups have also tested their predictions in the laboratory. The Severin group's efforts in this area began with a 2003 study in which the behavior of libraries undergoing self-selection was modeled [78]. Three types of libraries were considered (Fig. 1.25): type "A", in which one building block undergoes assembly into a library of compounds with variable stoichiometry; type "B", in which multiple building blocks form a library with fixed stoichiometry; and type "C", in which multiple building blocks undergo assembly into a library of variable stoichiometry. Calculations of steady-state concentrations for libraries of types "B" and "C" led to two somewhat surprising predictions. First, the selection of a structure containing a high percentage of one building block would drive formation of structures containing the other building blocks, and second, the selection processes (self or otherwise) naturally favor the production of heteroassemblies.

---

[3] Other contemporary reports supplemented experimental data with limited theoretical analysis; as the detailed simulation of library behavior was not the primary focus, they will not be discussed here.

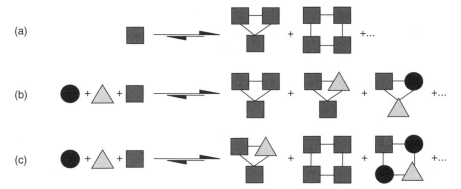

(a)

(b)

(c)

Figure 1.25 Library types examined computationally and experimentally by the Severin group.

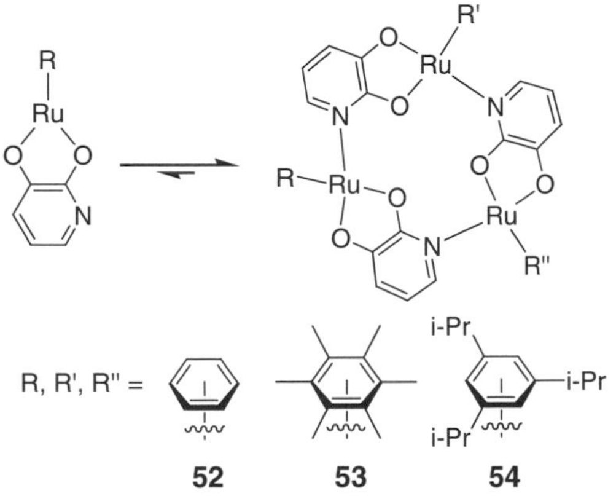

Figure 1.26 Experimental validation of predicted DCL behavior with metal-lomacrocycles.

Experimental confirmation of this predicted behavior was obtained via the examination of DCLs of trinuclear metallamacrocycles formed from ruthenium pyridonate complexes (Fig. 1.26). After first forming the individual homotrimers (i.e., **52–52–52**), these were mixed in various combinations and allowed to undergo exchange. As predicted, the thermodynamically most stable species was not always the most strongly selected.

In a subsequent paper, Severin expanded on these observations by defining boundary conditions under which DCLs can indeed provide the highest amplification of the most stable (or highest affinity) members [79]. The program Gepasi [80] was used to simulate the behavior of several different library types, including A, B, and C from the previous work as well as a new type, dubbed B*, in which all members of the library have a common subunit. Focusing in particular on target-induced adaptation, Severin found once again that equilibrium library distribution did not necessarily correspond to amplification of the tightest binder. However, this situation could be avoided by working with a relatively low concentration of target. Alternatively, Severin also suggested carrying out the selection in a "virtual" mode (in which the concentrations of library monomers outweigh concentrations of assemblies) as an alternative. Such libraries have been reported by Eliseev [81] and Lehn [82] as providing very large amplification factors for selected compounds.

In a pair of papers published in 2004 and 2005, Corbett, Otto, and Sanders described similar theoretical analyses of relatively simple DCLs, as well as the development and testing of a new tool for simulating DCL behavior, appropriately dubbed DCLSim [83]. The authors described the potential for DCL selections to yield something other than the "fittest" binder as the "tendency of DCLs to maximize the binding interactions in the entire library" [84]. As in the cases described by Severin, this tendency could be combated by careful choice of initial library conditions.

In more recent efforts, the Otto group has reported on the development and use of a related program, dubbed DCLFit, for converting experimentally observed product distributions from selection experiments into estimates of binding constants [85]. Experimental analysis of a number of different guest-templated libraries indicated a strong correlation between amplification factors and binding constants, consistent with the predictions of DCLSim and DCLFit [86]. Corbett, Sanders, and Otto have also noted that DCL experiments combined with simulation constitute an intriguing chemical approach to the study of complex systems [87]. We can anticipate that this "systems chemistry" will prove to be a vibrant field of research in the future. In the near term, such simulations are an important component of validating the DCC concept, as well as serving as critical guides to experimental design.

1.7. Conclusions

While the field of DCC is fairly young, it has extensive roots in other areas of endeavor, particularly in "origins of life" research and in self-assembly strategies from the supramolecular chemistry community. DCC has rapidly built an impressive and diverse set of applications, library types, and selection strategies. However, further research into new reversible reactions is needed in order to expand the range of structural types accessible to DCC. Efforts to simulate the behavior of DCLs are having a clear impact on the design of selection experiments, and we can anticipate that such studies will continue to represent valuable contributions to the field.

References

1. Bartlett, J. M. S.; Stirling, D. A short history of the polymerase chain reaction. *Methods Mol. Biol.* **2003**, *226*, 3–6.
2. Ellington, A. D.; Szostak, J. W. In vitro selection of RNA molecules that bind specific ligands. *Nature* **1990**, *346*, 818–822.
3. Tuerk, C.; Gold, L. Systematic evolution of ligands by exponential enrichment: RNA ligands to bacteriophage T4 DNA polymerase. *Science* **1990**, *249*, 505–510.
4. Smith, G. P.; Petrenko, V. A. Phage display. *Chem. Rev.* **1997**, *97*, 391–410.
5. Miller, S. L; Orgel, L. E. *The origins of life on earth*. Prentice Hall, Inc., Englewood Cliffs, NJ, 1974.
6. Joyce, G. F.; Visser, G. M.; van Boeckel, C. A. A.; van Boom, J. H.; Orgel, L. E.; van Westrenen, J. *Nature* **1984**, *310*, 602.
7. Goodwin, J. T.; Lynn, D. G. Template-directed synthesis: Use of a reversible reaction. *J. Am. Chem. Soc.* **1992**, *114*, 9197–9198.
8. Zhan, Z.-Y. J.; Lynn, D. G. Chemical amplification through template-directed synthesis. *J. Am. Chem. Soc.* **1997**, *119*, 12420–12421.
9. Miller, B. L.; Bonner, W. A. Enantioselective autocatalysis. III. Configurational and conformational studies on a 1,4-benzodiazepinooxazole derivative. *Orig. Life Evol. Biosph.* **1995**, *25*, 539–547.
10. Dhawan, B.; Chen, S. I.; Gutsche, C. D. Calixarenes. 19. Studies of the formation of calixarenes via condensation of p-alkylphenols and formaldehyde. *Makromol. Chem.* **1987**, *188*, 921–950.
11. Fouquey, C.; Lehn, J.-M.; Levelut, A.-M. Molecular recognition directed self-assembly of supramolecular liquid crystalline polymers from complementary chiral components. *Adv. Mater.* **1990**, *2*, 254–257.
12. Krämer, R.; Lehn, J.-M.; Marquis-Rigault, A. Self-recognition in helicate self-assembly: Spontaneous formation of helical metal complexes from mixtures of ligands and metal ions. *Proc. Natl. Acad. Sci. USA* **1993**, *90*, 5394–5398.
13. Zhao, D.; Moore, J. S. Reversible polymerization driven by folding. *J. Am. Chem. Soc.* **2002**, *124*, 9996–9997.
14. Oh, K.; Jeong, K.-S.; Moore, J. S. *m*-Phenylene ethynylene sequences joined by imine linkages: Dynamic covalent oligomers. *J. Org. Chem.* **2003**, *68*, 8397–8403.

15. Zhao, D.; Moore, J. S. Nucleation–elongation polymerization under imbalanced stoichiometry. *J. Am. Chem. Soc.* **2003**, *125*, 16294–16299.

16. Zhao, D.; Moore, J. S. Folding-driven reversible polymerization of oligo (*m*-phenylene ethynylene) imines: Solvent and starter sequence studies. *Macromolecules* **2003**, *36*, 2712–2720.

17. Shaked, Z.; Szajewski, R. P.; Whitesides, G. M. Rates of thiol–disulfide interchange reactions involving proteins and kinetic measurements of thiol pKa values. *Biochemistry* **1980**, *19*, 4156–4166.

18. Hioki, H.; Still, W. C. Chemical evolution: A model system that selects and amplifies a receptor for the tripeptide (D)Pro(L)Val(D)Val. *J. Org. Chem.* **1998**, *63*, 904–905.

19. Still, W. C. Discovery of sequence-selective peptide binding by synthetic receptors using encoded combinatorial libraries. *Acc. Chem. Res.* **1996**, *29*, 155–163.

20. McNaughton, B. R.; Gareiss, P. C.; Miller, B. L. Identification of a selective small-molecule ligand for HIV-1 frameshift-inducing stem-loop RNA from an 11,325 member resin bound dynamoic combinatorial library. *J. Am. Chem. Soc.*, **2007**, *129*, 11306–11307.

21. Gareiss, P. C.; Sobczak, K.; McNaughton, B. R.; Thornton, C. A.; Miller, B. L. Dynamic combinatorial selection of small molecules capable of inhibiting the (CUG) repeat RNA–MBNL1 interaction in vitro: Discovery of lead compounds targeting myotonic dystrophy (DM1). *J. Am. Chem. Soc.*, **2008**, *130*, 16524–16261.

22. Ludlow, R. F.; Otto, S. Two-vial, LC-MS identification of ephedrine receptors from a solution-phase dynamic combinatorial library of over 9000 components. *J. Am. Chem. Soc.* **2008**, *130*, 12218–12219.

23. Huc, I.; Lehn, J. M. Virtual combinatorial libraries: Dynamic generation of molecular and supramolecular diversity by self-assembly. *Proc. Natl. Acad. Sci. U.S.A.* **1997**, *94*, 2106–2110.

24. Wessjohann, L. A.; Rivera, D. G.; León, F. Freezing imine exchange in dynamic combinatorial libraries with Ugi reactions: Versatile access to templated macrocycles. *Org. Lett.* **2007**, *9*, 4733–4736.

25. Xu, S.; Giuseppone, N. Self-duplicating amplification in a dynamic combinatorial library. *J. Am. Chem. Soc.* **2008**, *130*, 1826–1827.

26. Ziach, K.; Jurczak, J. Controlling and measuring the equilibration of dynamic combinatorial libraries of imines. *Org. Lett.* **2008**, *10*, 5159–5162.

27. Bunyapaiboonsri, T.; Ramström, H.; Ramström, O.; Haiech, J.; Lehn, J.-M. Generation of bis-cationic heterocyclic inhibitors of *Bacillus subtilis* HPr kinase/phosphatase from a ditopic dynamic combinatorial library. *J. Med. Chem.* **2003**, *46*, 5803–5811.

28. Nazarpack-Kandlousy, N.; Nelen, M. I.; Goral, V.; Eliseev, A. V. Synthesis and mass spectrometry studies of branched oxime ether libraries. Mapping the substitution motif via linker stability and fragmentation pattern. *J. Org. Chem.* **2002**, *67*, 59–65.

29. Wipf, P.; Mahler, S. G.; Okumura, K. Metathesis reactions of pyrazolot-riazinones generate dynamic combinatorial libraries. *Org. Lett.* **2005**, *7*, 4483–4486.

30. Cordes, E. H.; Bull, H. G. Mechanism and catalysis for hydrolysis of acetals, ketals, and ortho esters. *Chem. Rev.* **1974**, *74*, 581–603.

31. Bone, R.; Cullis, P.; Wolfenden, R. Solvent effects on equilibria of addition of nucleophiles to acetaldehyde and the hydrophilic character of diols. *J. Am. Chem. Soc.* **1983**, *105*, 1339–1343.

32. Sorensen, P.; Jencks, W. Acid- and base-catalyzed decomposition of acetaldehyde hydrate and hemiacetals in aqueous solution. *J. Am. Chem. Soc.* **1987**, *109*, 4675–4690.

33. Fuchs, B.; Nelson, A.; Star, A.; Stoddart, J. F.; Vidal, S. Amplification of dynamic chiral crown ether complexes during cyclic acetal formation. *Angew. Chem. Int. Ed.* **2003**, *42*, 4220–4224.

34. Cacciapaglia, R.; Di Stefano, S.; Mandolini, L. Metathesis reaction of formaldehyde acetals: An easy entry into the dynamic covalent chemistry of cyclophane formation. *J. Am. Chem. Soc.* **2005**, *127*, 13666–13671.

35. Berkovich-Berger; Lemcoff, N. G. Facile acetal dynamic combinatorial library. *Chem. Commun.* **2008**, 1686–1688.

36. Brady, P. A.; Sanders, J. K. M. Thermodynamically controlled cyclisation and interconversion of oligocholates: Metal ion templated 'living' macrolac-tonisation. *J. Chem. Soc. Perkin I* **1997**, 3237–3253.

37. Rowan, S. J.; Hamilton, D. G.; Brady, P. A.; Sanders, J. K. M. Automated recognition, sorting, and covalent self-assembly by predisposed building blocks in a mixture. *J. Am. Chem. Soc.* **1997**, *119*, 2578–2579.

38. Rowan, S. J.; Sanders, J. K. M. Macrocycles derived from cinchona alkaloids: A thermodynamic vs. kinetic study. *J. Org. Chem.* **1998**, *63*, 1536–1546.

39. Woll, M. G.; Gellman, S. H. Backbone thioester exchange: A new approach to evaluating higher order structural stability in polypeptides. *J. Am. Chem. Soc.* **2004**, *126*, 11172–11174.

40. Larsson, R.; Ramström, O. Dynamic combinatorial thiolester libraries for efficient catalytic self-screening of hydrolase substrates. *Eur. J. Org. Chem.* **2006**, *2006*, 285–291.

41. LeClaire, J.; Vial, L.; Otto, S.; Sanders, J. K. M. Expanding diversity in dynamic combinatorial libraries: Simultaneous exchange of disulfide and thioester linkages. *Chem. Commun.* **2005**, 1959–1961.

42. Heumann, A.; Réglier, M. The stereochemistry of palladium-catalysed cyclisation reactions. Part B: Addition to pi-allyl intermediates. *Tetrahedron* **1995**, *51*, 975–1015.

43. Kaiser, G.; Sanders, J. K. M. Synthesis under reversible conditions of cyclic porphyrin dimers using palladium-catalysed allyl transesterification. *Chem. Commun.* **2000**, 1763–1764.

44. Miller, B. L.; Klekota, B. Apparatus and methods describing a two-chambered reaction vessel for ligand-affinity target identification using a combinatorial library. US Patent 6,599,754.

45. Uozumi, Y.; Suzuka, T. pi-Allylic sulfonylation in water with amphiphilic resin-supported palladium–phosphine complexes. *Synthesis* **2008**, 1960–1964.

46. Yang, X. W.; Gong, B. Template-assisted cross olefin metathesis. *Angew. Chem. Int. Ed.* **2005**, *44*, 1352–1356.

47. Chen, Y.-Y.; He, W.-Y.; Wu, Y.; Niu, C.-Q.; Liu, G. Dynamic selection of novel vancomycin N-terminal derivatives by resin-bound reversed D-Ala-D-Ala. *J. Comb. Chem.* **2008**, *10*, 914–922.

48. McNaughton, B. R.; Bucholtz, K. M.; Camaano-Moure, A.; Miller, B. L. Self-selection in olefin cross metathesis: The effect of remote functionality. *Org. Lett.* **2005**, *7*, 733–736.

49. Ven Gerven, P. C. M.; Elemans, J. A. A. W.; Gerritsen, J. W.; Speller, S.; Nolte, R. J. M.; Rowan, A. E. Dynamic combinatorial olefin metathesis: Templated synthesis of porphyrin boxes. *Chem. Commun.* **2005**, 3535–3537.

50. Zhang, W.; Moore, J. S. Arylene ethynylene macrocycles prepared by pre-cipitation-driven alkyne metathesis. *J. Am. Chem. Soc.* **2004**, *126*, 12796.

51. Zhang, W.; Moore, J. S. Reaction pathways leading to arylene ethynylene macrocycles via alkyne metathesis. *J. Am. Chem. Soc.* **2005**, *127* 11863–11870.

52. Boul, P. J.; Reutenauer, P.; Lehn, J.-M. Reversible Diels–Alder reactions for the generation of dynamic combinatorial libraries. *Org. Lett.* **2005**, *7*, 15–18.

53. Bennes, R. M.; Philp, D. Probing selectivity in recognition-mediated dynamic covalent processes. *Org. Lett.* **2006**, *8*, 3651–3654.

54. Eliseev, A. V.; Nelen, M. I. Use of molecular recognition to drive chemical evolution. 1. Controlling the composition of an equilibrating mixture of simple arginine receptors. *J. Am. Chem. Soc.* **1997**, *119*, 1147–1148.

55. Ingerman, L. A.; Waters, M. L. Photoswitchable dynamic combinatorial libraries: Coupling azobenzene photoisomerization with hydrazone exchange. *J. Org. Chem.* **2009**, *74*, 111–117.

56. Bamberger, E. Kondensationsproducte aus *o*-amino-benzaldehyd, nebst einem anhang: Zur darstellung diese aldehyds. *Chem. Ber.* **1927**, *60*, 314–319.

57. Seidel, F.; Dick, W. Über die anhydroverbindungen des o-amino-benzaldehyds (2. Mitteilung). *Chem. Ber.* **1927**, *60*, 2018–2023.

58. Klekota, B.; Hammond, M. H.; Miller, B. L. Generation of novel DNA-binding compounds by selection and amplification from self-assembled combinatorial libraries. *Tetrahedron Lett.* **1997**, *38*, 8639–8643.

59. Klekota, B.; Miller, B. L. Selection of DNA-binding compounds *via* multi-stage molecular evolution. *Tetrahedron* **1999**, *55*, 11687 (Combinatorial Chemistry Symposium in Print).

60. Sakai, S.; Shigemasa, Y.; Sasaki, T. A self-adjusting carbohydrate ligand for GalNAc specific lectins. *Tetrahedron Lett.* **1997**, *38*, 8145–8148.

61. Buryak, A.; Severin, K. Easy to optimize: Dynamic combinatorial libraries of metal–dye complexes as flexible sensors for tripeptides. *J. Comb. Chem.* **2006**, *8*, 540–543.

62. Buryak, A.; Severin, K. Dynamic combinatorial libraries of dye complexes as sensors. *Angew. Chem. Int. Ed.* **2005**, *44*, 7935–7938.

63. Buryak, A.; Pozdnoukhov, A.; Severin, K. Pattern-based sensing of nucleotides in aqueous solution with a multicomponent indicator displacement assay. *Chem. Commun.* **2007**, 2366–2368.

64. Buryak, A.; Zaubitzer, F.; Pozdnoukhov, A.; Severin, K. Indicator displacement assays as molecular timers. *J. Am. Chem. Soc.* **2008**, *130*, 11260–11261.

65. Mal, P.; Schultz, D.; Beyeh, K.; Rissanen, K.; Nitschke, J. R. An unlockable–relockable iron cage by subcomponent self-assembly. *Angew. Chem. Int. Ed.* **2008**, *47*, 8297–8301.

66. Eldred, S. E.; Stone, D. A.; Gellman, S. H.; Stahl, S. S. Catalytic transamidation under moderate conditions. *J. Am. Chem. Soc.* **2003**, *125*, 3422–3423.

67. Hoerter, J. M.; Otte, K. M.; Gellman, S. H.; Cui, Q.; Stahl, S. S. Discovery and mechanistic study of Al(III)-catalyzed transamidation of tertiary amides. *J. Am. Chem. Soc.* **2008**, *130*, 647–654.

68. Goral, V.; Nelen, M. I.; Eliseev, A. V.; Lehn, J. M. Double-level "orthogonal" dynamic combinatorial libraries on transition metal template. *Proc. Natl. Acad. Sci. U.S.A.* **2001**, *98*, 1347–1352.

69. Mikami, M.; Shinkai, S. Synthesis of helical polymers by polycondensation of diboronic acid and chiral tetrols. *Chem. Lett.* **1995**, 603–604.

70. Nakazawa, I.; Suda, S.; Masuda, M.; Asai, M.; Shimizu, T. pH-dependent reversible polymers formed from cyclic sugar- and aromatic boronic acid-based bolaamphiphiles. *Chem. Commun.* **2000**, 881–882.

71. Christinat, N.; Scopelliti, R.; Severin, K. Multicomponent assembly of boronic acid based macrocycles and cages. *Angew. Chem. Int. Ed.* **2008**, *47*, 1848–1852.

72. Lipinski, C. A.; Lombardo, F.; Dominy, B. W.; Feeney, P. J. Experimental and computational approaches to estimate solubility and permeability in drug discovery and development settings. *Adv. Drug Deliv. Rev.* **1997**, *23*, 3–25.

73. Epstein, D. M.; Choudhary, S.; Churchill, M. R.; Keil, K. M.; Eliseev, A. V.; Morrow, J. R. Chloroform-soluble Schiff-base Zn(II) or Cd(II) complexes from a dynamic combinatorial library. *Inorg. Chem.* **2001**, *40*, 1591–1596.

74. Choudhary, S.; Morrow, J. R. Dynamic acylhydrazone metal ion complex libraries: A mixed-ligand approach to increased selectivity in extraction. *Angew. Chem. Int. Ed.* **2002**, *41*, 4096–4098.

75. Pérez-Fernández, R.; Pittelkow, M.; Belenguer, A. M.; Sanders, J. K. M. Phase-transfer dynamic combinatorial chemistry. *Chem. Commun.* **2008**, 1738–1740.

76. Moore, J. S.; Zimmerman, N. W. 'Masterpiece' copolymer sequences by targeted equilibrium-shifting. *Org. Lett.* **2000**, *2*, 915–918.

77. Connors, K. A. The stability of cyclodextrin complexes in solution. *Chem. Rev.* **1997**, *97*, 1325–1358.

78. Grote, Z.; Scopelliti, R.; Severin, K. Adaptive behavior of dynamic combinatorial libraries generated by assembly of different building blocks. *Angew. Chem. Int. Ed.* **2003**, *42*, 3821–3825.

79. Severin, K. The advantage of being virtual – Target-induced adaptation and selection in dynamic combinatorial libraries. *Chem. Eur. J.* **2004**, *10*, 2565–2580.

80. Mendes, P. Biochemistry by numbers: Simulation of biochemical pathways with Gepasi 3. *Trends Biochem. Sci.* **1997**, *22*, 361–363.

81. Hochgürtel, M.; Kroth, H. Piecha, D.; Hofmann, M. W.; Nicolau, C.; Krause, S.; Schaaf, O.; Sonnenmoser, G.; Eliseev, A. V. Target-induced formation of neuraminidase inhibitors from in vitro virtual combinatorial libraries. *Proc. Natl. Acad. Sci. U.S.A.* **2002**, *99*, 3382–3387.

82. Nitschke, J. R.; Lehn, J.-M. Self-organization by selection: Generation of a metallosupramolecular grid architecture by selection of components in a dynamic library of ligands. *Proc. Natl. Acad. Sci. U.S.A.* **2003**, *100*, 11970–11974.

83. Corbett, P. T.; Otto, S.; Sanders, J. K. M. Correlation between host–guest binding and host amplification in simulated dynamic combinatorial libraries. *Chem. Eur. J.* **2004**, *10*, 3139–3143.

84. Corbett, P. T.; Sanders, J. K. M.; Otto, S. Competition between receptors in dynamic combinatorial libraries: Amplification of the fittest? *J. Am. Chem. Soc.* **2005**, *127*, 9390–9392.

85. Ludlow, R. F.; Liu, J.; Li, H.; Roberts, S. L.; Sanders, J. K. M.; Otto, S. Host–guest binding constants can be estimated directly from the product distributions of dynamic combinatorial libraries. *Angew. Chem. Int. Ed.* **2007**, *46*, 5762–5764.

86. Corbett, P. T.; Sanders, J. K. M.; Otto, S. Exploring the relation between amplification and binding in dynamic combinatorial libraries of macrocyclic synthetic receptors in water. *Chem. Eur. J.* **2008**, *14*, 2153–2166.

87. Corbett, P. T.; Sanders, J. K. M.; Otto, S. Systems chemistry: Pattern formation in random dynamic combinatorial libraries. *Angew. Chem. Int. Ed.* **2007**, *46*, 8858–8861.

Chapter 2

Protein-Directed Dynamic Combinatorial Chemistry

Michael F. Greaney and Venugopal T. Bhat

2.1. Introduction

The use of proteins to control the evolution of dynamic combinatorial libraries (DCLs) has been central to the development of the field, since the first publications in the area appeared in the late 1990s [1]. The protein-directed DCL concept is a simple one: A DCL of compounds is designed such that it can equilibrate in the presence of a protein. Since the library population distribution is under thermodynamic control, stabilization of one member through selective binding to the protein is expected to amplify that species at the expense of other (nonbinding) species, generating hit structures that can be identified through analysis of the DCL population distribution [2]. Protein-directed dynamic combinatorial chemistry (DCC) thus provides a method for studying, discovering, and ranking novel protein-binding ligands, concepts fundamental to medicinal chemistry [3–6]. In these terms, the DCC process bridges the gap between chemical synthesis of drug candidates and their biological binding assay, meshing the two operations into a single process whereby the structure of the biological target directs the assembly of its own best inhibitor *in situ*.

Protein targets are challenging templates to work with in DCC. Their sensitivity to pH, temperature, and chemical reagents places significant restraint on the synthetic chemistry that can be successfully used in

Dynamic Combinatorial Chemistry, edited by Benjamin L. Miller
Copyright © 2010 John Wiley & Sons, Inc.

protein-templated DCC. The DCLs must be assembled under essentially physiological conditions, and it is no surprise that many of the bond formations used to create the reversible DCC linkages, such as $S-S$ or $C=N$ bond formation, are fundamental to biological chemistry.

Directly interfacing a DCL with a protein template and analyzing the amplified distribution of small molecules has been termed as *adaptive DCL*, and represents the simplest and most powerful realization of the DCC concept. In many cases it has not been possible to follow the adaptive principle with protein targets, alternative DCL systems requiring the design of that can approach this ideal case in terms of effective amplification of best binding DCL components. Indeed, the difficulty of implementing and analyzing reversible chemistry in the presence of proteins has stimulated some highly innovative ideas that have been influential in the development of the DCC field at large.

This chapter will cover all DCC systems that use proteins to influence the population distribution of DCLs, with the exception of Ramström's coupled DCC systems that are discussed separately in Chapter 6. The DCL systems under discussion are loosely grouped according to the chemical reactions used to set up the DCLs. The chemistry used to make protein-directed DCLs is in many ways fundamental, as it defines the conditions in which the protein can or cannot operate as a molecular trap for the DCL. Compatibility of the various synthetic reactions with protein targets will establish whether a genuine adaptive DCL can be constructed, or whether alternative approaches are needed to separate the fragile biomolecule from the DCL chemistry.

2.2. C=N Bond Formation

The first explicit treatment of the DCC concept using proteins was reported by Lehn and Huc in 1997 [1,7].[1] In a touchstone paper for the development of the DCC area in general, they synthesized an imine DCL to probe the active site of the enzyme carbonic anhydrase (CA). The composition of

---

[1] References 1 and 7 describe 'virtual combinatorial libraries' (VCLs). The choice of the word virtual is expounded upon by the author in Reference 7, whereby DCC is contrasted with VCLs. This elegant bifurcation of terminology has not been widely taken up in the literature, possibly due to confusing overlap with terminology describing VCLs in the area of *in silico* drug design.

the library is shown in Fig. 2.1, consisting of four amines **a–d** and three aldehydes **1–3**. The choice of structure was based on a number of factors.

- All components were selected to present structural features similar to those of known CA inhibitors. The three aldehydes, for example, contain a *para*-substituted sulfonamide motif, which is known to bind the Zn (II) ion in the enzyme's active site with submicromolar activity.
- The functional groups creating the imine linkage were selected to have comparable activity, that is, aromatic aldehydes plus amines of similar nucleophilicity. This approach seeks to avoid large differences in reactivity that might excessively bias the library composition.
- All the components had to display activity in the UV-visible spectrum, which would aid detection and analysis by HPLC.

The DCL was composed using a 15-fold excess of amines with respect to individual aldehydes, at pH 6 in the presence of 3-fold excess of NaB-H_3CN. The large excess of amines ensured pseudo first-order behaviour with respect to the aldehyde, and limited crossreactivity with nucelophilic amino acid residues on the protein surface. The reducing agent is necessary as the imine products in the DCL are too unstable for analysis/isolation. Slow reduction produces sturdy amines that are easily analyzed and highly tractable in medicinal chemistry. The drawback, however, is that the imine molecules taking part in the crucial molecular recognition events that dictate the templating and amplification process in the DCL are not represented in the final analysis. A further practical consideration concerns the reduction of the starting aldehydes as a side reaction, which erodes the aldehyde concentration in the DCL and produces benzylic alcohols **1′–3′** as side products.

The DCL was first composed in the absence of the CA template, a so-called blank DCL. Equilibration was complete in 24 hours, with each of the expected imine reduction products being observed in the HPLC trace (Fig. 2.2). Reconstitution of the DCL in the presence of a stoichiometric quantity of enzyme afforded the second trace. Equilibration was significantly retarded in the presence of the protein, 2 weeks being necessary for a dynamic equilibrium to be realized. A thermal denaturation step preceded HPLC analysis.

Close inspection of the templated library revealed a number of changes relative to the blank with respect to each set of amines. Lower overall yields were generally observed for all reduction products in the presence of the enzyme, and the relative composition of amines derived from **1** and **2**

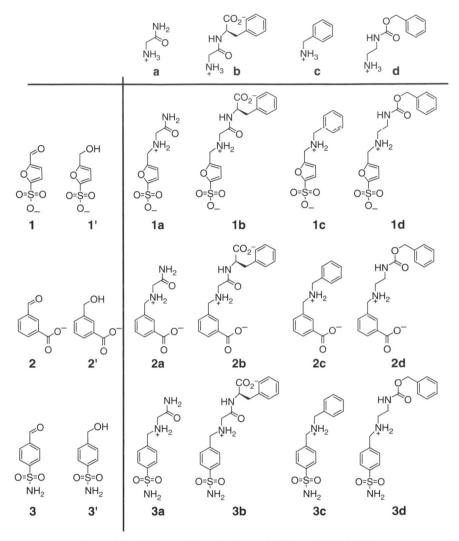

Figure 2.1   Lehn and Huc's imine DCL for CA inhibition.

was essentially unchanged. The amines created from aldehyde **3**, however, had undergone a change in abundance in the presence of CA, with the benzylamine derivative **3c** being amplified at the expense of **3a** and **3d**. Amine **3b** maintained the same relative concentration.

Further DCL experiments were carried out focussing solely on aldehyde **3** and the four amines, enabling a binding series to be established. An important control experiment was carried out, whereby the DCL was constituted in the presence of hexyl 4-sulfamoylbenzoate, a known CA

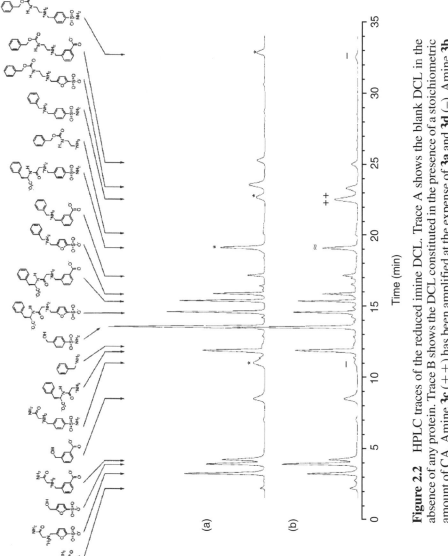

Figure 2.2 HPLC traces of the reduced imine DCL. Trace A shows the blank DCL in the absence of any protein. Trace B shows the DCL constituted in the presence of a stoichiometric amount of CA. Amine **3c** (++) has been amplified at the expense of **3a** and **3d** (−). Amine **3b** is relatively unchanged (≈). Reproduced from Reference 1 with permission of the National Academy of Sciences, USA. Copyright (1997) National Academy of Sciences, U.S.A.

47

inhibitor. No amplifications were observed, implicating the active site of CA as being responsible for the templation. The structure of the imine precursor to the best binder **3c** is very similar to the known high affinity CA ligand 4-sulfamoylbenzoic acid benzylamide (K_d= 1.1 nM).

Lehn and Huc's imine DCC study proved the principle of DCC on biological targets, and formed the blueprint for how DCLs could be constructed for the interrogation of biological targets. The same research group have examined acylhydrazone formation as a method for DCL generation with a system directed toward inhibiting acetylcholine esterase (ACE) (Fig. 2.3) [8]. Acylhydrazone formation was introduced by the Sanders group as a powerful method for DCC in abiotic systems [9]. The reaction features readily available aldehydes and hydrazide components, and crucially, forms acyl hydrazone products that are robust and readily amenable

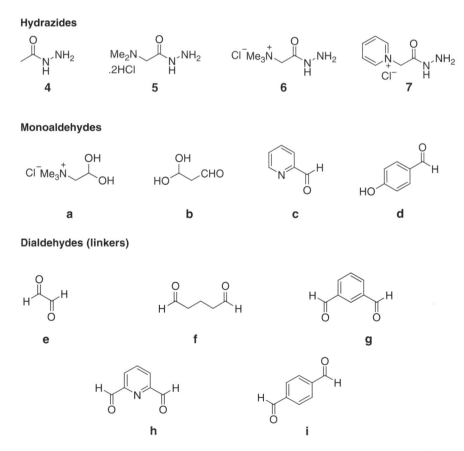

Figure 2.3   Hydrazide and aldehyde components for acyl hydrazone DCL.

to analysis. A potential barrier to the use of acyl hydrazone formation in DCC with biological targets is the acidic pH (ca. 4) required for efficient equilibration, which will denature most protein targets. This proved to be the case for ACE, and a "pre-equilibrated" DCC approach was taken whereby the DCL is generated in the absence of any target, frozen, and then exposed to the target in a conventional enzyme assay in a second discrete step. The separation of DCL synthesis and templating processes allows a greater range of reversible reactions to be used, with no restrictions due to the delicate nature of the protein target. However, it removes the adaptive and amplification processes that define DCC, with the second phase akin to a conventional screen of a combinatorial mixture library. As with that approach, a deconvolution strategy is required to identify hit compounds from the mixture.

As proof of principle, Lehn and coworkers individually synthesized all acyl hydrazone combinations from the 13 DCL building blocks and measured their inhibition of acetylthiocholine hydrolysis by ACE in a standard assay. They then established a *dynamic deconvolution* approach whereby the pre-equilibrated DCL containing all members is prepared, frozen, and assayed. Thirteen sublibraries were then prepared containing all components minus one hydrazide or aldehyde component, and assayed. Active components in the DCL were quickly identified by an increase in ACE activity, observed in sublibraries missing either hydrazide **7** or dialdehyde **i**, pointing to the bis-acyl hydrazone **7–i–7** as the most likely active constituent. This was in line with the individual assay data recorded earlier; resynthesis of this compound characterized it as a low nanomolar inhibitor of the enzyme.

The same pre-equilibrated acyl hydrazone DCL approach was taken by Lehn and coworkers in a medicinal chemistry study to discover cationic inhibitors of *Bacillus subtilis* HPr kinase [10]. The DCL was designed around the structure **8**, a lead inhibitor discovered from an in-house library screen (Fig. 2.4). The library contained 16 hydrazides containing nitrogen heterocycles, two monoaldehydes and 3 dialdehydes, producing a theoretical 440-membered DCL on acid equilibration. Using the dynamic deconvolution strategy previously established, construction and subsequent assay of 21 sublibraries clearly identified dialdehyde **10** and hydrazide **9** as being responsible for most inhibitory activity. Resynthesis of the corresponding bis-acyl hydrazone **11** allowed a more detailed study of inhibition, revealing an IC_{50} value of 18 μM.

Simple imine formation was used in the formation of large DCLs directed toward inhibition of the enzyme neuraminidase by Eliseev and coworkers [11]. Neuraminidase is involved in the propagation of the influenza virus

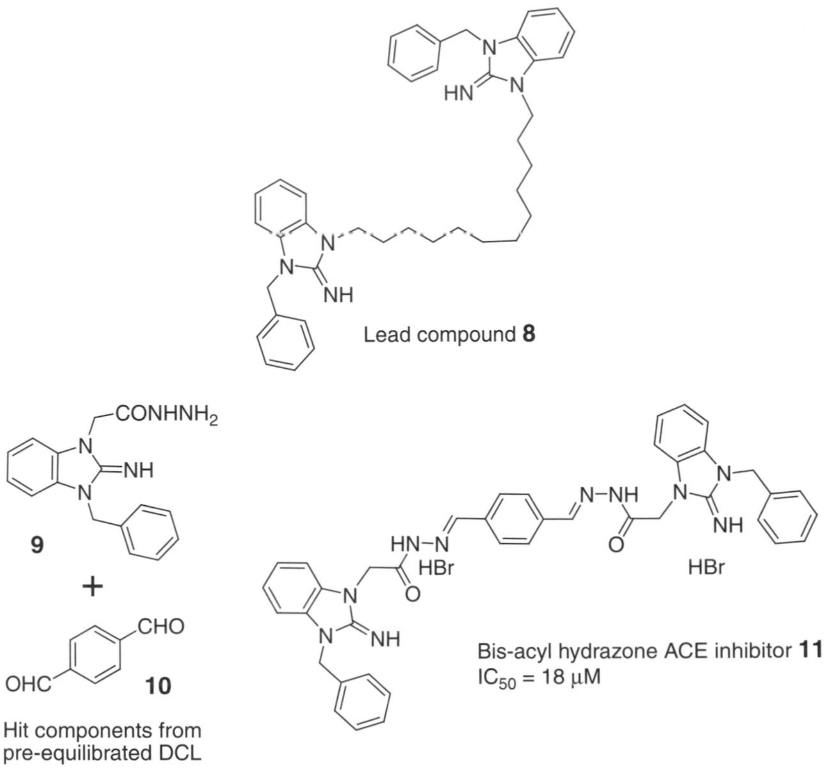

Figure 2.4 Key acyl hydrazone DCL components for kinase inhibition.

and is a major target in influenza drugs such as Oseltamivir Tamiflu, **12**. The related diamine compound **13** was used as the scaffold for DCL formation, whereby aldehydes chosen from a pool of 41 commercially available compounds were added in the presence of sodium cyanoborohydride (Scheme 2.1). As with Lehn and Huc's canonical work, the imine DCC products are thus reduced to amines *in situ*, which are analyzed by HPLC. The construction of the DCL is noteworthy for two reasons. First, it is theoretically very large. If 20 aldehydes are used, over 40,000 components are possible at equilibrium if one counts hemi-aminal structures, a reasonable inclusion as they could be observed by ^{1}H NMR of sample DCLs. Second, use of a single, divalent amine component means that the entire DCL can in theory be amplified to a single compound. This contrasts with earlier work (*vide supra*) where the $n \times n$ ($n > 1$) stoichiometry precludes such unilateral amplifications.

An interesting analytical method was adopted for the DCL. Given the potential size of the library, Eliseev and coworkers reasoned that it would not be possible, or even desirable, to attempt a detailed characterization of all components present at equilibrium. Instead, the library was run under

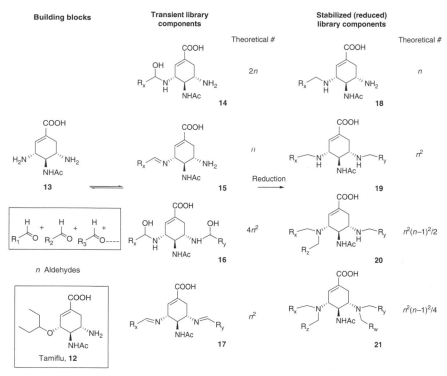

Scheme 2.1   Imine DCL built from diamine scaffold 13 and *n* aldehydes.

conditions where the transient components were present at very low concentrations, and NH_4BH_3CN reduction was slow in a given time frame. In the absence of target, no DCL members can be detected. In the presence of neuraminidase, however, amplification of the best binders affords detectable amounts of compound that may be identified by HPLC-MS. In the event, the DCL strongly amplified the three adducts **18a**, **18b**, and **18c** of the monosubstituted class, with **18c** being amplified an estimated 120 times relative to its concentration in the blank DCL (Fig. 2.5).

Control experiments verified that the neuraminidase active site was responsible for controlling DCL evolution. First, substitution of neuraminidase with bovine serum albumin as the target produced no amplification at all. Second, reconstituting the DCL in the presence of excess zanamivir, a commercially available potent inhibitor of neuraminidase, likewise gave results identical to the blank DCL in the absence of any target.

Given the absence of well-characterized blank DCLs, it was not possible to say with certainty that the amplified compounds were the result of genuine adaptive behaviour of the DCL at thermodynamic equilibrium,

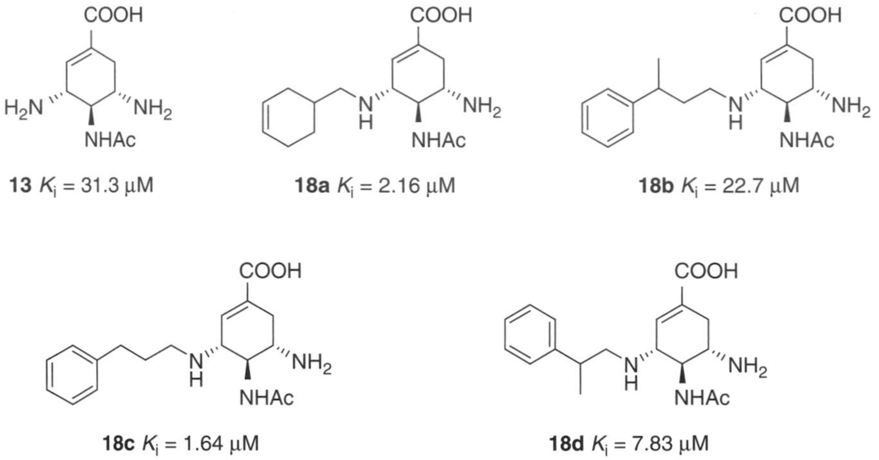
COOH
H_2N
NH_2
NHAc
13 K_i = 31.3 μM
COOH
N
H
NH_2
NHAc
18a K_i = 2.16 μM
COOH
N
H
NH_2
NHAc
18b K_i = 22.7 μM
COOH
N
H
NH_2
NHAc
18c K_i = 1.64 μM
COOH
N
H
NH_2
NHAc
18d K_i = 7.83 μM

Figure 2.5 K_i values for amplified compounds from imine DCLs.

or whether kinetic acceleration accounted for the production of the amplified compounds, or a mixture of both. However, given the well-known reversible formation of imines, precedented in the DCC context, it is a fair assumption that a DCC effect was operating.

Resynthesis of a selection of the hit compounds **18** and biological assay revealed significantly increased activity for **18a** and **18c** relative to the scaffold **13**, validating the DCL approach for discovery of neuraminidase inhibitors. The results also revealed weaknesses inherent to the imine DCL system, namely, the generation of false positive and negative results. Amine **18b** was amplified in the DCL, but is not a particularly good binder, displaying only slightly improved activity over the scaffold—a false positive. Likewise, amine **18d** was not amplified at all in the DCLs, but was resynthesized and assayed as a somewhat better inhibitor—a false negative. These results stem from the separation of the molecular recognition events that control amplification in the DCL, namely binding the transient imines to the protein's active site, and those that control binding of the product amines.

Eliseev and coworkers subsequently reported a follow-up study on their DCC system using ketones in place of aldehydes and the same Tamiflu analog **13** as the scaffold [12].

Beau has applied the DCC concept to a number of carbohydrate systems [13–16]. Carbohydrate-binding proteins often exhibit weak ligand interactions with millimolar dissociation constants, a quite different scenario to the enzyme–small molecule DCLs discussed thus far. The protein target initially chosen for study was hen egg-white lysozyme (HEWL), a glycosidase known to bind *N*-acetyl-D-glucosamine (D-GlcNAc) with

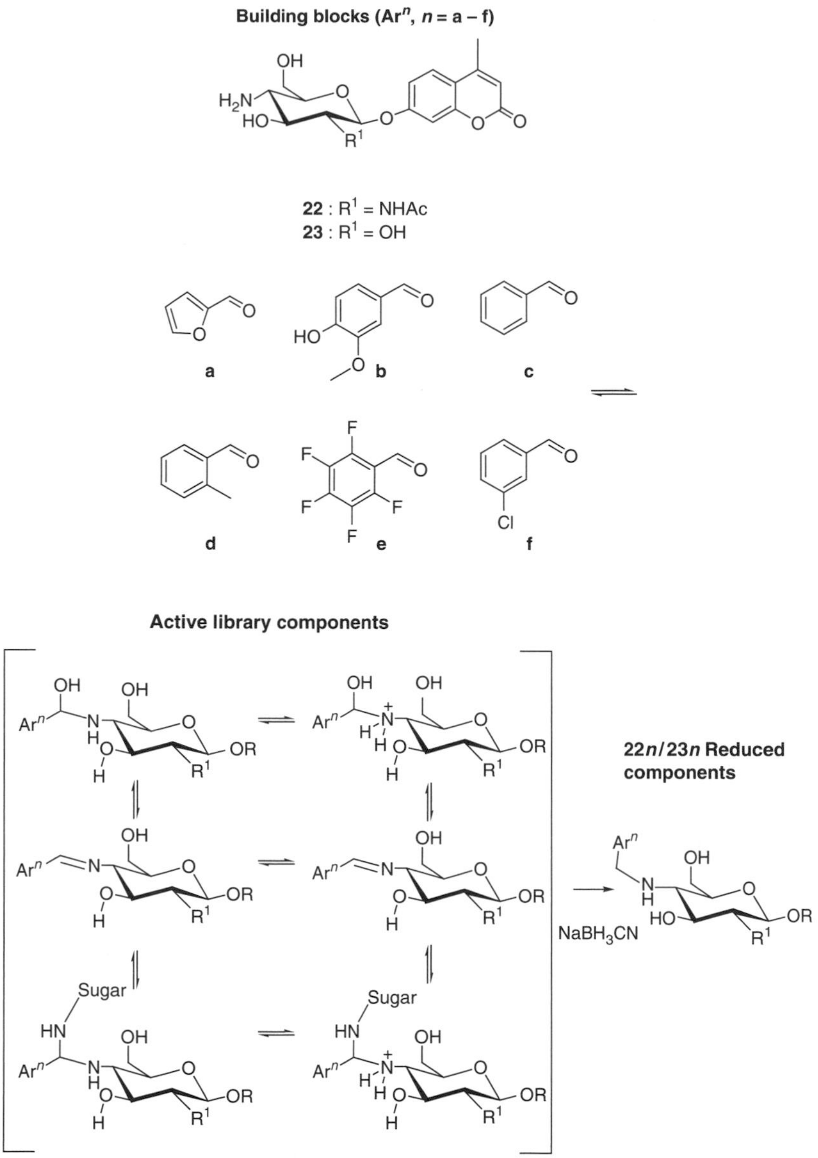

Building blocks (Ar^n, n = a – f)

22 : R^1 = NHAc
23 : R^1 = OH

a   b   c

d   e   f

Active library components

22n/23n Reduced components

NaBH$_3$CN

Scheme 2.2   DCL of D-GlcNAc imines.

low millimolar affinity [13]. The imine bond was chosen as the reversible linkage: The two D-GlcNAc and D-Glc compounds **22** and **23** were equilibrated with equimolar amounts of six benzaldehydes (**a–f**) at pH 6.2 in the presence of sodium cyanoborohydride (Scheme 2.2).

HPLC analysis, aided by the 4-methylumbelliferyl chromophore appended to the carbohydrate anomeric position, identified all 12 of the expected amines. Equilibration in the presence of a stoichiometric quantity of HEWL led to a small amplification of two components **22b** and **22f**, which could be augmented by using an excess of protein as the DCL template (Fig. 2.6). As a control, the DCL was constructed in the presence of chitotriose, a good HEWL inhibitor, and no amplification could be observed.

The most amplified amine, **22b**, was resynthesized and assayed against the rate of HEWL lysis of *Micrococcus lysodeikticus*. Competitive inhibition kinetics was observed, with a recorded K_i value of 0.6 mM. The amplified amine **22b** is thus around 100 times more active than the starting D-GlcNAc scaffold.

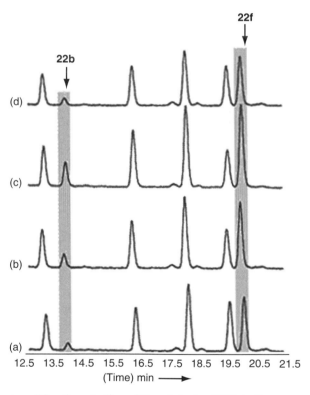

Figure 2.6 Amplification studies of Beau's imine DCL in the presence of HWL. (**a**) no HWL, (**b**) 1 equiv. of HWL, (**c**) 3 equiv. of HWL, and (**d**) 1 equiv. of HWL and 3 equiv. of chitotriose. Reproduced from Reference 13 with permission of Wiley-VCH Verlag GmbH & Co. KGaA. Copyright Wiley-VCH Verlag GmbH & Co. KGaA.

Having established that DCC could be effective for the interrogation of weak protein–carbohydrate interactions in a glycolytic enzyme, the Beau group next examined glycosyltransferases (GTs) as DCC targets [14]. Rational design of GT inhibitors is difficult due to the absence of GT crystal structures, allied with a complex active site function that weakly binds the sugar acceptor, donor, and metal catalytic center. The DCC approach suffers no such limitations, as structural information of the target is not needed to effectively amplify good binding components from DCLs. One drawback, however, lies in the requirement for stoichiometric amounts of the target protein necessary for measurable amplifications. This drawback becomes apparent if GTs are used as targets, as they are typically available in very small amounts.

The unavailability of stoichiometric quantities of GT protein target prompted Beau to formulate an interesting analysis of the imine DCL idea. It seems reasonable to suppose that the products of imine DCC, reduced amines, should share the binding characteristics of the parent imines that undergo the molecular amplification *in situ*. This proves to be the case when stoichiometric amounts of protein can be used. In cases where protein supply is limited and substoichiometric amounts must be contemplated, competitive binding of the amine products to the target may erode amplification over time. It would in fact be better if the amine products bound significantly more weakly than the imine DCL members, removing the competitive inhibition problem and permitting constant amplification of the reduced amine products. This scenario was realized using α-1,3-galactosyltransferase (α1,3GalT) to direct a carbohydrate-based imine DCL.

Library design is illustrated in Fig. 2.7, and is based upon UDP-galactose **24**, the natural substrate of α1,3GalT. The DCL was designed around three classes of components: nucleoside aldehydes **25–29**, diamine linkers **a–g**, and D-galactose aldehyde **30**. Using one nucleoside building block at a time, small DCLs were assembled using a subset of the seven diamine linkers plus carbohydrate **30**. Aldehyde and amine concentrations (82 and 542 μM) were significantly higher than the GT target, at 1.6 μM. Best results were obtained for a DCL containing aldehydes **25** and **30** and diamines **a**, **b**, **e**, and **f**, which produced the reduced monoamine products as DCL products. Double amination, where both the nucleoside and galactosyl aldehyde had condensed with a diamine, was not observed. Running the DCL in the presence of the GT amplified all monoamine products, with **25e** and **25f** in particular being amplified relative to **25a**, **25b**, and **30f**. A control experiment with BSA was negative, implicating molecular recognition by the GT target as being responsible for amplification. Surprisingly, a second control using the natural substrate UDP was also negative, suggesting that the imine DCL components are not bound to the GT active site.

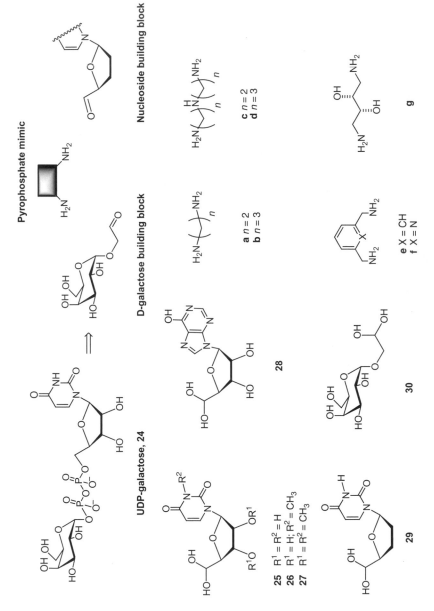

Figure 2.7 Building blocks for imine DCL templated by α 1,3GalT.

The inhibitory activity of amplified amines **25e** and **25f** was measured relative to the amides **25'e** and **25'f** to shed light on the initial reasoning that $K_{amine} \ll K_{imine}$ would be necessary for successful amplification, the amides representing "geometric" mimics of the corresponding imine DCL components. Inhibition assays established IC_{50} values of > 5 mM for both amines (UDP-Gal IC_{50}= 73 μM), with the amides showing values of 1.1 and 0.4 mM, respectively (Fig. 2.8).

Beau and coworkers extended their work on galactosyltransferases by examining β-1,4-galactosyltransferase (β-*galactosyltransferase*) enzyme as a target for their carbohydrate DCL. The β-*galactosyltransferase* forms the glycosidic bond with inversion of configuration, whereas the α-GT does so with retention. Templating the DCL in Fig. 2.7 with the β-GT produced quite different results, with only modest amplifications being observed for amines **25a** and **25b**.

Congreve and coworkers at Astex pharmaceuticals have developed a DCC system that uses X-ray crystallography to observe ligand binding [17,18].[2] Using cyclin-dependant kinase 2 (CDK2) as the target, a heavily studied kinase in cancer research, a DCL was designed around the aryl hydrazine and isatin motifs **A** and **B** (Scheme 2.3).

The viability of hydrazone formation as the DCC reaction was first assessed by taking each of the 30 pairs of reactants in aqueous dimethylsulfoxide and monitoring their reaction by LC-MS. In each case the product hydrazone could be identified, verifying that the forward reaction was occurring successfully. The reverse reaction was not investigated. The hydrazones were then formed in the presence of CDK2 crystals, first as separate pairs of hydrazine and isatin reactants, then as mixtures. Following soaking of

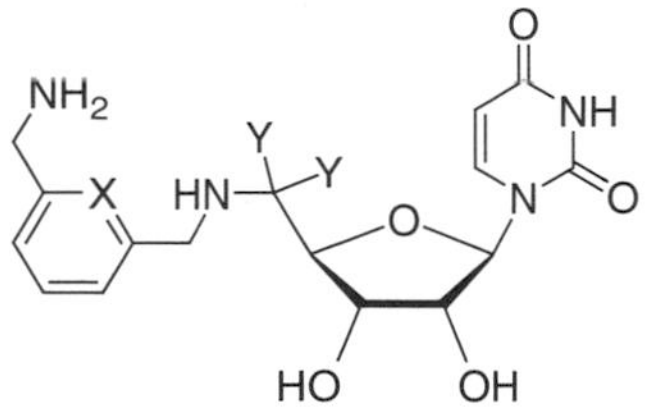

25e	X = CH, Y = H	IC_{50} = 5 mM
25f	X = N, Y = H	IC_{50} = 5 mM
25'e	X = CH, Y = (=O)	IC_{50} = 1.1 mM
25'f	X = N, Y = (=O)	IC_{50} = 0.4 mM

Figure 2.8   Assay data for DCL hit compounds.

[2] For an early example of using X-ray crystallography to identify binding events templated by a protein, see Reference 18.

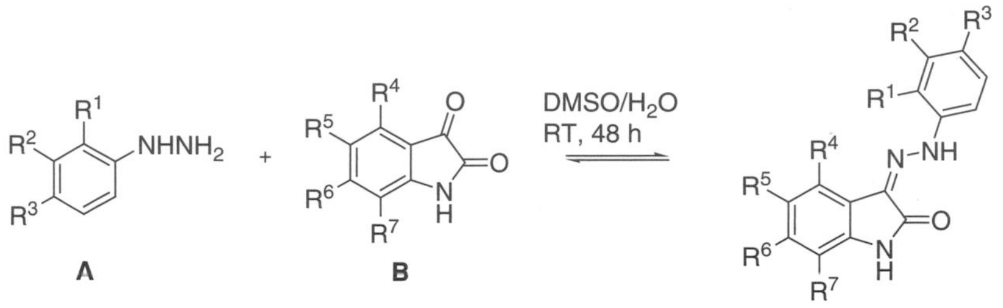

A31 - (R groups = H unless indicated otherwise)
A32 R^1 = Cl
A33 R^2 = Cl
A34 R^3 = Cl
A35 R^3 = SO$_2$NH$_2$
A36 R^1 = Cl, R^3 = SO$_2$Me

B37 R^5 = NO$_2$
B38 R^5 = Cl
B39 R^5 = CF$_3$
B40 R^7 = CF$_3$
B41 R^5 = OCF$_3$

Scheme 2.3 Hydrazone formation for DCC analyzed by X-ray crystallography.

the crystal in each of the combinations of **35A** with the isatins, X-ray crystallography and inspection of the resultant electron density maps showed ligand binding in four of the five combinations. Subsequent biological assay confirmed inhibitory activity at the 30 nM level. The one combination (**A35B41**) that did not show binding in the soaking experiment was found to be biologically inactive in a subsequent assay. The study was extended to a mixture of all 30 combinations using the six hydrazines and five isatins in a single soaking experiment. Comparison of electron density maps identified the potent hydrazone **A35B38** as a binder.

Given the very small amount of protein present in a single crystal relative to the concentrations of DCL components, it is clear that amplification in the usual sense cannot occur. It is conceivable, however, that the protein structure influences equilibrium distribution of hydrazones in microcosm within the crystal. Conversely, it is possible that any DCC equilibrium is irrelevant, and that the hydrazone binding being observed by X-ray diffraction is due to a diffusion equilibrium of essentially static components between solution and solid state.

2.3. Disulfide Bond Exchange

Disulfide bond formation was introduced into DCC as a powerful reaction for the construction of dynamic systems in the late 1990s in separate reports from the groups of Still [19], Sanders [20], and Lehn [21]. Given the fundamental role played by thiol oxidation in biology, it is no surprise that the reaction is highly compatible with protein targets. Disulfide exchange proceeds readily in water at mildly basic pH in the presence of oxygen and catalytic amounts of thiol; acidification turns off exchange. Unlike the imine linkage, disulfide bonds are sufficiently stable to undergo analysis and isolation from aqueous solutions. The reaction is also chemoselective, tolerating a number of functional groups. The biological compatibility of disulfide exchange may also be assumed to be good, despite the presence of cysteine residues in proteins. Disulfide linkages are frequently encountered within the interior of proteins, and are unlikely to be accessed by small-molecule thiols at low concentration.

The first example of disulfide bond formation using a protein target was reported by Lehn and Ramström using a carbohydrate DCL [21]. The lectin concanavalin A (Con A) is a well-studied carbohydrate-binding protein that is specific for the branched trimannoside unit (Man)$_3$, **42**. The design of the library was predicated on the assumption that the central mannoside moiety in (Man)$_3$ acts as a linker, with the two peripheral mannosides being mainly responsible for binding interactions to the lectin. Accordingly, mannosides were derivatized with a phenylamido chromophore appended with short-chain alkyl thiols (Scheme 2.4). The thiols would form disulfides under DCL conditions, with the disulfide chain acting as a linker between the two carbohydrate head-groups. This design strategy aims for an isoenergetic DCL, whereby the equilibrium constant for all library members undergoing the disulfide exchange reaction is close to unity, and the rate of reaction between components is likewise similar. Significant disparities in reaction rate between reacting functional groups will result in biased DCLs that do not express the full range of product structures at equilibrium. This feature is desirable, as it will maximize the structural diversity that can compete for receptor binding in the library. The disulfide linkage in **43** is thus isolated from the carbohydrate-binding elements, which should produce an isoenergetic DCL without significant kinetic disparity.

The DCL formed smoothly using four disulfides **43a**, **43d**, **43e**, and **43f** as starting components equilibrated at pH 7.4 for 2 weeks in the presence of dithiothreitol (DTT) (Fig. 2.9). All 10 of the expected ditopic carbohydrate combinations could be observed by HPLC. In a novel departure from existing DCL systems of the time, the Con A target was attached to sepharose beads, the first such example of an immobilized protein being used in DCC. On addition of the protein, amplification could be assessed first by quantifying components that had been removed from the solution DCL, and second by applying an acidic quench to the DCL and washing bound components off the resin beads.

Distinction was made between the addition of Con A from the beginning of DCL equilibration, and afterwards (the pre-equilibrated approach previously applied by Lehn to acylhydrazone DCLs) [8]. In the first approach, a

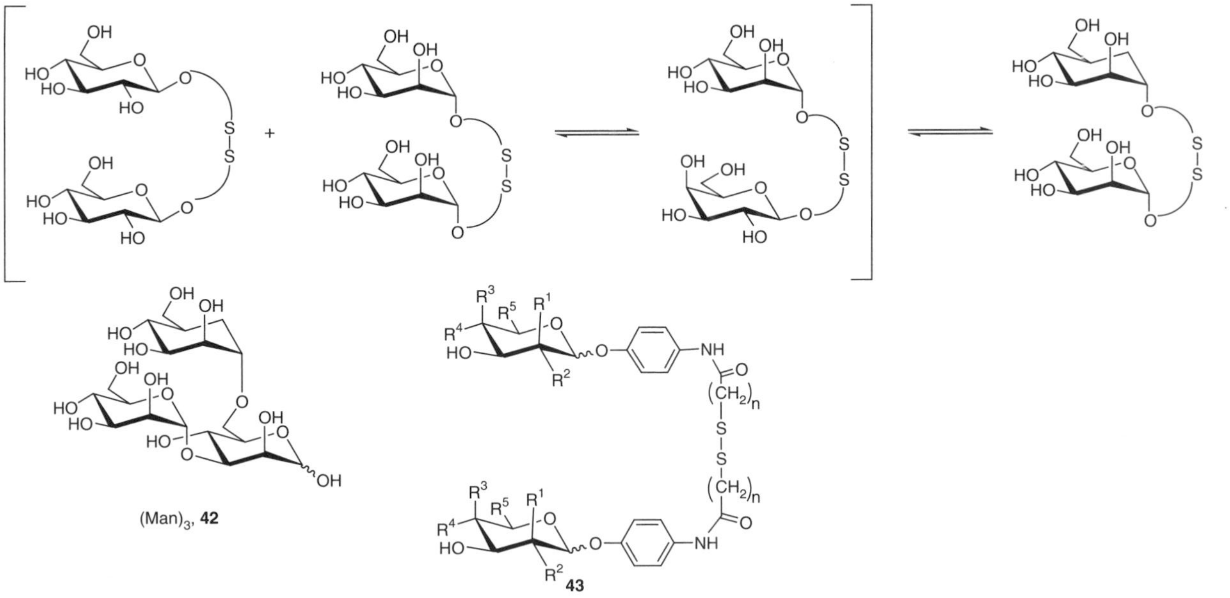

Compound	α/β	R¹	R²	R³	R⁴	R⁵	n
43a (Man/Man)	α	OH	H	H	OH	CH₂OH	3
43b (GalC₂/GalC₂)	β	H	OH	OH	H	CH₂OH	2
43c (GalC₃/GalC₃)	β	H	OH	OH	H	CH₂OH	3
43d (Glc/Glc)	β	H	OH	H	OH	CH₂OH	2
43e (Ara/Ara)	β	H	OH	OH	H	H	2
43f (Xyl/Xyl)	β	H	H	H	OH	H	2

Man = D-mannose; GalC₂ = D-galactose; GalC₃ = D-galactose, n = 2; GalC₃ = D-galactose, n = 3; Glc = D-glucose; Ara = L-arabinose; Xyl = D-xylose

Scheme 2.4 Disulfide exchange DCL for interrogation of Con A.

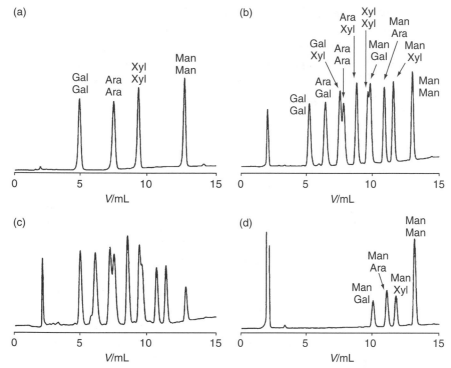

Figure 2.9 LC analysis of disulfide DCLs. (**a**) Starting disulfides **43a**, **43d**, **43e**, and **43f**, (**b**) equilibrated DCL (no target), (**c**) equilibrated DCL in the presence of Con A, and (**d**) eluate of bound species from immobilized Con A. Reproduced from Reference 21 with permission of Wiley-VCH Verlag GmbH & Co. KGaA. Copyright Wiley-VCH Verlag GmbH & Co. KGaA.

reduction in the concentration of several D-mannose-containing dimers in the DCL solution was observed, most notably for the Man–Man homodimer. Acidification and elution of the beads identified the dimers as being bound to Con A (Fig. 2.9d). Adding the receptor upon pre-equilibration afforded the similar selectivities. Analogous preference for the natural mannose substrates was observed in the larger library featuring all six disulfide building blocks. Analysis of the DCL solution at equilibrium was hampered by HPLC-resolution issues. Quenching and elution, however, clearly showed the Man dimers bound to the immobilized protein.

While the amplification of the DCL by Con A was relatively weak, as expected for such a low affinity binding system, the system demonstrated the effectiveness of mild disulfide bond exchange when applied to DCC directed by a protein target. Lehn subsequently reported a second DCC investigation into the Con A binding site involving sugars linked via acyl hydrazone conversion [22]. The acidic conditions of the reaction dictated a pre-equilibration/deconvolution approach.

Hunter and Waltho have examined disulfide exchange in peptidic DCLs directed towards binding of the calcium transducer calmodulin (CaM) [23]. The DCL was based upon a known binding motif for CaM consisting of two hydrophobic peptides connected by a flexible linker—ideally suited to a DCC investigation. The library components are shown in Fig. 2.10, and consist of cystine dimers containing hydrophobic amino acid residues.

The DCL was created under standard disulfide exchange conditions at pH 7.5. After equilibration for 48 hours, each of the expected 15 disulfide products could be observed in the library using LC-MS analysis (Fig. 2.11). The library was then equilibrated in the presence of CaM, followed by a centrifugation filtration step to separate protein/bound components from free components in solution. Analysis of the filtrate was complicated by the filtration membrane affecting the composition of the library. The bound components, however, provided meaningful results. Denaturation and filtration afforded a mixture of all peptides that had bound to CaM in the course of the DCL. HPLC analysis indicated significant amplification of dimer **cc** (80%) and a small amplification of dimer **ec** (10%). Resynthesis of these two components and binding assay established K_d values of 10 and 210 µM, respectively, over two orders of magnitude lower than the binding constant of the monovalent thiol NSH.

A similar approach to a DCL of bivalent ligands linked by flexible disulfide spacers was taken by Danieli and coworkers [24]. Their library consisted of the antitumour agents thiocolchicine and podophyllotoxin, derivatized to form disulfide-linker homo- and heterodimers (Scheme 2.5). The components could be effectively exchanged in the presence of catalytic thiol in organic solvents.

Unfortunately, the low aqueous solubility of the DCL components precluded templating studies with the intended target microtubulin. As proof of principle, the proteins albumin and subtilisin were substituted as organic solvent tolerant alternatives. Starting from the homodimeric compounds **45** and **48**, the amount of heterodimer **51** formed was shown to be influenced by the presence of the proteins.

Researchers at Sunesis pharmaceuticals have developed a fragment-based drug discovery method termed tethering [25]. The approach, which is illustrated in Scheme 2.6, shares a number of features with DCC. Whereas protein-directed DCLs equilibrate small molecules via disulfide formation, say, in the presence of a protein that acts as a thermodynamic trap, tethering uses a cysteine residue on the protein surface to reversibly capture small-molecule thiol fragments from solution. Tethering is designed

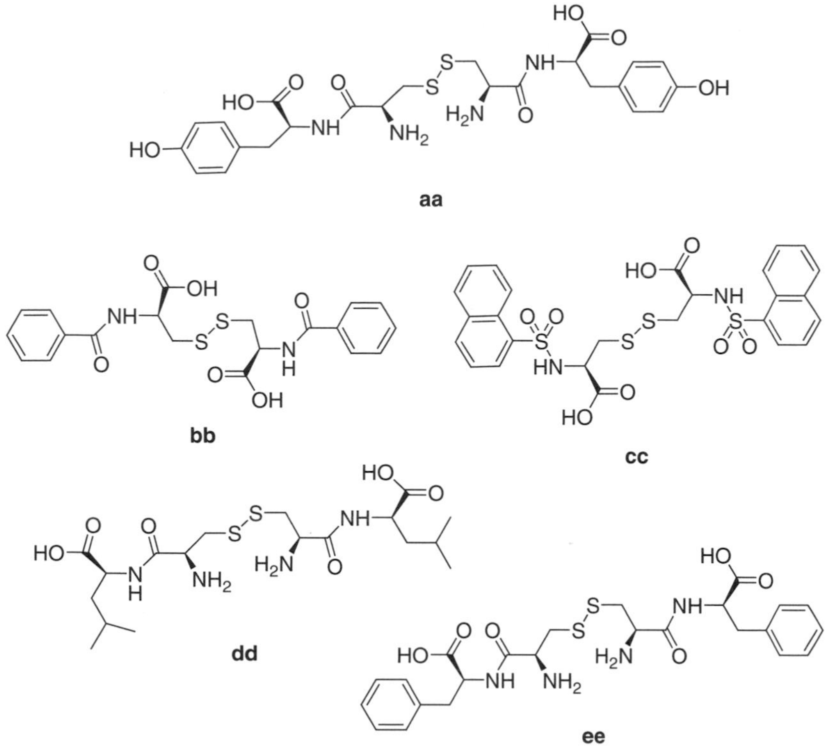

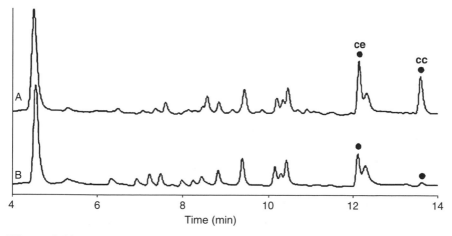

Figure 2.10   Cystine dimer components for calmodulin DCL.

Figure 2.11   LC analysis of cystine DCL. Trace A shows the DCL templated with CaM, and trace B shows the blank DCL. The amplified peaks (·) are dipeptides **ce** and **cc**. Reproduced from Reference 23 with permission of Wiley-VCH Verlag GmbH & Co. KGaA. Copyright Wiley-VCH Verlag GmbH & Co. KGaA.

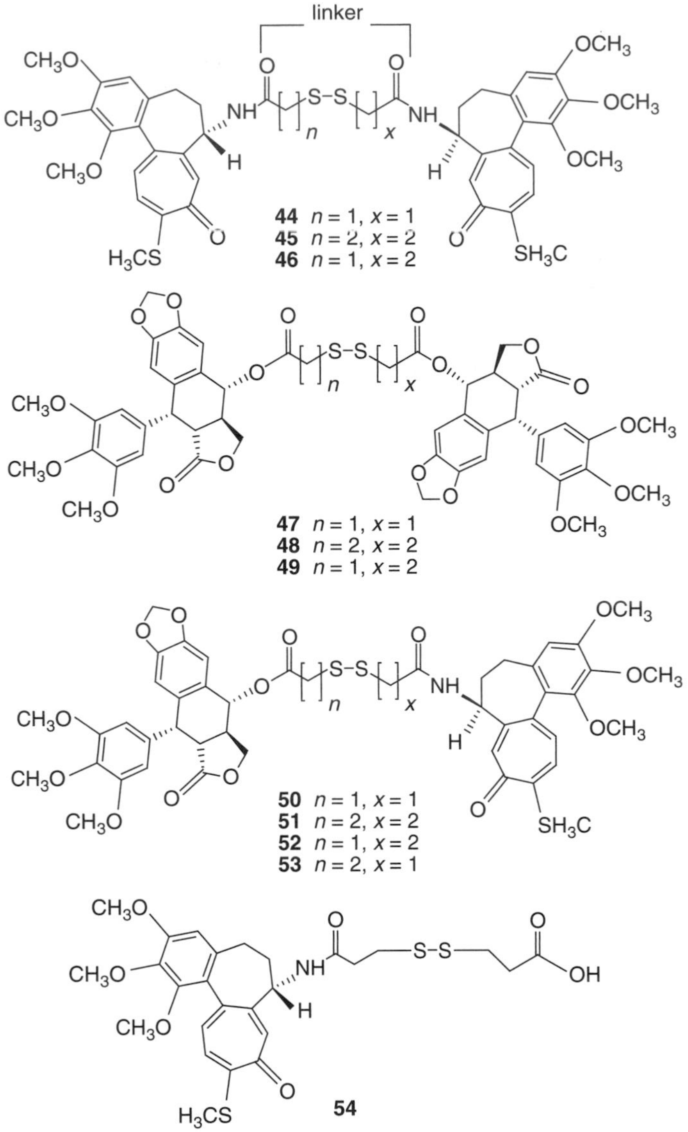

Scheme 2.5 Disulfide DCL created from thiocolchicine and podophyllotoxin.

to identify small, weakly binding ligands that can be used as fragments to construct larger more potent drug candidates. Clearly, cysteine residues close to the active site of interest are required, and if they are not present naturally, then they must be introduced using site-directed mutagenesis. While this technique is a standard technology in protein chemistry, the knowledge of where to place the requisite thiol can only be gained with

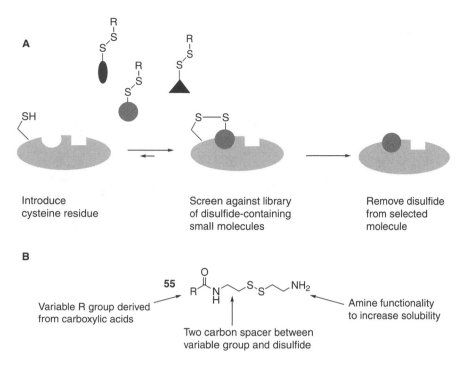

Scheme 2.6 The tethering concept.

access to structural information. Tethering has been applied with success to a number of targets [26–30], one of which will be discussed here by way of example [31].

The approach was first exemplified using the enzyme thymidylate synthase (TS), a well-described anticancer target that contains an active site cysteine. The small-molecule components were designed around the general structure **55**, containing a small fragment part derived from a carboxylic acid that could be linked to a disulfide containing a terminal amine. The terminal amine is present to increase solubility. Libraries were constructed using 8–15 disulfides, each at 0.2 mM, in the presence of 15-μM protein. The disulfide bonds were exchanged under standard conditions and then subjected to LC-MS analysis with the hit components covalently bound to the protein. The compounds are identified by simple subtraction of the known protein mass from the observed protein-tethered complex mass. This analysis depends on each tethering experiment featuring fragments of distinct molecular mass, preferably differentiated by 10 atomic mass units, and also on the reasonable assumption that the ionization profile of each tethered protein-complex is essentially the same.

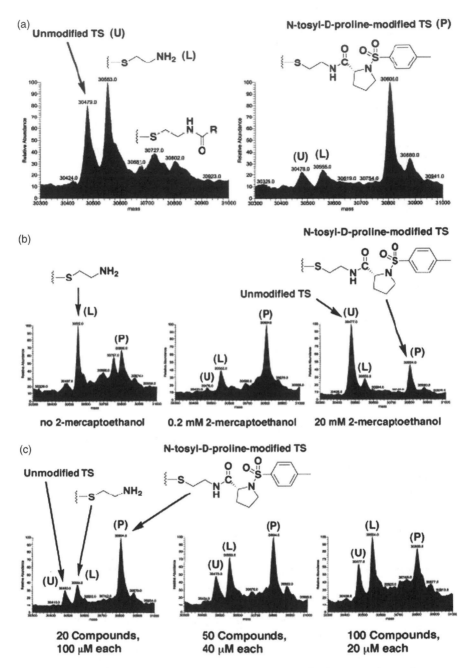

Figure 2.12   MS data of tethered complexes. (**a**) Tethering experiment between TS and 10 disulfides equilibrated for 1 hour, (**b**) tethering with varying concentrations of 2-mercaptoethanol, and (**c**) tethering with varying pool size of disulfides. Reproduced from Reference 31 with permission of the National Academy of Sciences, USA. Copyright (2000) National Academy of Sciences, U.S.A.

The results are illustrated in Fig. 2.12. For tethering experiments that produce no binders, the deconvoluted MS shows peaks for the unmodified protein (U) along with TS tethered to 2-aminoethanethiol (L), which is common to all library members. Hit compounds are clearly observed, such as the *N*-tosyl-D-proline fragment (P) shown in Fig. 2.12a. The tethering experiments could be put on a firm thermodynamic footing by varying the amount of catalytic thiol (2-mercaptoethanol) added in the disulfide equilibration conditions. In the absence of thiol, only moderate selectivity for the hit fragment P over L was observed, as the system does not feature thermodynamic reversibility. Addition of thiol sets up DCL conditions and results in the strong selection of hit fragment P. A particular strength of the approach proved to be its scalability, a key challenge in DCC. The *N*-tosyl-proline fragment could be clearly selected from tethering experiments containing 100 compounds (Fig. 2.12c). Given the rapidity of disulfide exchange, plus MS analysis, the technique is thus in principle amenable to high-throughput screening.

For the system at hand, tethering experiments involving 1200 compounds produced a structure-activity profile around the proline fragment P, supplemented with crystallographic studies of TS with the fragment bound to the active site. Importantly, the conformation of fragment P in the covalently linked, tethered complex was found to be very similar to a noncovalent complex between *N*-tosyl-D-proline and TS. This finding demonstrates that the covalent linkage is not affecting the binding mode of the active fragment. Fragment P was subsequently developed by appending the side chain of the TS cofactor methylenetetrahydrofolate to the tosyl group in P to produce inhibitors of vastly improved affinity.

2.4. Enzymatic Methods

The use of enzymes to catalyze reversible reactions has proven to be an effective strategy for DCC. Enzymes work under physiological conditions (by definition), are reversible, and can also be applied to a variety of C–C and C–X bond-forming reactions. Venton and coworkers reported the first example of an enzyme-catalyzed process being used in a DCC context [32]. As their work preceded the codification of DCC in the literature, it contains little of the vocabulary that has come to define the field. It does, however, correspond perfectly with the conceptual framework of DCC, and has been widely cited as an influential early example of the DCC idea.

The enzymatic reaction chosen for the DCL was protease-catalyzed amide bond synthesis/hydrolysis. This fundamental transformation is difficult to replicate in a reversible manner using chemical methods, although there has been recent progress in the area [33–35]. Taking the tripeptide YGG and the dipeptide FL as starting points, incubation with the broad-spectrum protease thermolysin produced over 15 short peptide sequences formed from hydrolysis and synthesis of the amino acid building blocks present in the substrates. The pentapeptide YGGFL was identified as being formed in 0.5% yield from the reaction. In a similar reaction, the dipeptides VA and AL were incubated with thermolysin, followed by a digestion with cathepsin C. Seven out of the possible nine dipeptides could be identified, proving that the protease effectively scrambles the amino acid sequence of the starting materials.

The protease DCL was then studied in the presence of a receptor fibrinogen. Fibrinogen is a protein involved in the blood-clotting process and is conveniently available in milligram quantities from commercial sources. The known binding tripeptide motif GPR was incubated with thermolysin and bovine serum albumin (BSA) hydrolysates under the scrambling conditions, with the fibrinogen template separated by a dialysis membrane. Comparison with control experiments identified three peptides as being synthesized by thermolysin and amplified by fibrinogen: GPRL, GPRF, and DKPDNF. The two tetrapeptides were found to bind fibrinogen weakly ($K_a = 10^{-4}M^{-1}$), an affinity twofold lower than GPR.

Flitsch and Turner reported the generation of a sialic acid DCL using an aldolase enzyme [36,37]. The DCL design is shown in Scheme 2.7, and is based on the cleavage of sialic acid (**58a**) to N-acetylmannosamine (**56a**) and sodium pyruvate (**57**), catalyzed by an aldolase enzyme.

A small DCL was set up using D-mannose (**56b**) and D-lyxose (**56c**) as additional substrates. HPLC of the aldol products established equilibration taking place over 16 hours, with reversibility being proven by re-equilibration

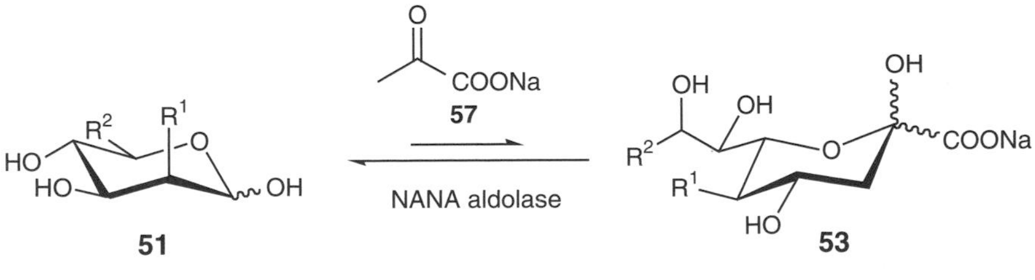

56a R^1 = NHAc, R^2 = CH$_2$OH
56b R^1 = OH, R^2 = CH$_2$OH
56c R^1= OH, R^2 = H

58a R^1 = NHAc, R^2 = CH$_2$OH
58b R^1 = OH, R^2 = CH$_2$OH
58c R^1 = OH, R^2 = H

Scheme 2.7   Aldol reaction of ManNAc analogues and sodium pyruvate to produce sialic acid, catalyzed by N-acetylneuraminic acid (NANA) aldolase.

experiments. The target for the DCL was the plant lectin wheat germ agglutinin (WGA), known as a weak binder of sialic acid. The templated DCL showed clear amplification for the natural substrate sialic acid (**58a**), at the expense of the adduct **58b** (Fig. 2.13). After 160 hours, **58a** comprised 40% of the total aldol peak area compared to 22% in the blank control experiment. The concentration of adduct **58c** stayed relatively unchanged, whereas that of **58b** fell from 31% to 6%. Microcalorimetry established the expected weak binding affinity of **58a** to WGA of 172 M^{-1}, whereas no detectable binding could be measured for **58b**.

Gleason and Kazlauskas have introduced the concept of pseudodynamic combinatorial chemistry [38,39], which combines an irreversible synthesis of library components with an irreversible destruction step. The idea is illustrated in Scheme 2.8 for a pseudo-DCL of peptides designed to inhibit

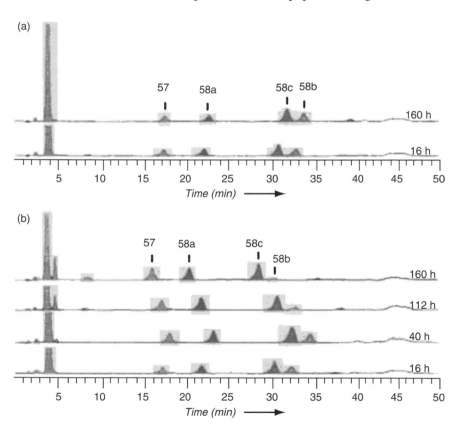

Figure 2.13 HPLC analysis of aldolase-mediated DCL templated by wheat germ agglutinin (WGA. (**a**) Blank DCL and (**b**) DCL composition in the presence of WGA. Reproduced from Reference 36 with permission of Wiley-VCH Verlag GmbH & Co. KGaA. Copyright Wiley-VCH Verlag GmbH & Co. KGaA.

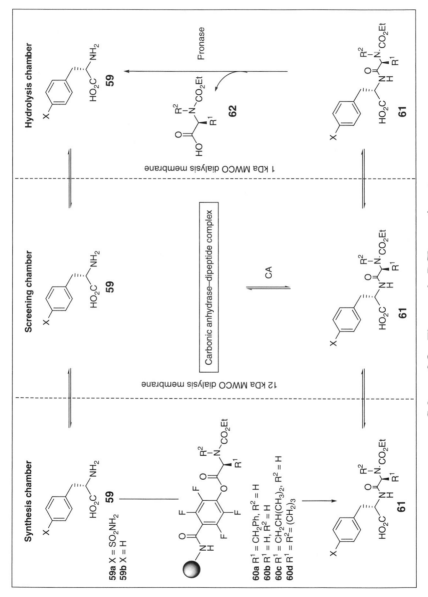

Scheme 2.8 The pseudo-DCL experiment.

CA. The reaction system is compartmentalized, a key feature. In practice this separation is achieved using two dialysis bags suspended in a surrounding solution. In the first compartment, the synthesis chamber, dipeptide binders are created through the irreversible addition of amino acids (**59**) to the immobilized active ester **60**. These dipeptides (**61**) may then diffuse into the screening chamber, which contains the target CA, and a binding equilibrium is established. Further diffusion into a third compartment, the destruction chamber, hydrolyzes the dipeptides back to the amino acids using a large excess of the broad-spectrum protease pronase. The amino acid building blocks can then migrate back to the synthesis chamber to be reused as starting materials for reaction with **60** (which is periodically restocked). The crux of the process lies in the fact that the higher affinity library members will be protected from destruction by binding to CA, initiating an amplification process. While reversible chemical reactions are thus not a feature of the pseudo-DCL approach, it conforms perfectly to the DCC credo of amplifying the best binders at the expense of the poor ones.

The pseudo-DCL at hand employed two amino acid building blocks **59a** and **59b** plus four different immobilized active esters **60a–d**. A total of eight dipeptides were assayed against CA to establish a rank order of affinity—the four phenylalanine derivatives do not bind CA and are present as negative controls. A critical requirement of the pseudo-DCL is that large differences in rate are not encountered in either synthesis or destruction reactions. Control experiments established that this proved to be the case; a large excess of pronase ensured that any natural selectivity of the enzyme was over-ridden in the destruction chamber. In the synthesis chamber, the active esters **60a–d** were found to be reasonably indiscriminate for coupling **59a**, producing significant amounts of all four of the expected dipeptides in a control experiment.

After some optimization of the cycle time (time between replenishment of **60**), effective amplification of the best binding peptide could be achieved. After seven cycles (112 hours), a 100-fold selectivity for the dipeptide derived from **60a** and **60d** could be observed over the next best binder, a remarkable result given that the ratio of binding constants is only 2.3:1. This selectivity is far higher than that typically observed in DCLs, and is based in part on the iterative nature of the experiment, an approach first demonstrated with success by Eliseev in abiotic dynamic systems [40,41].

2.5. Other Methods

Nicolaou and coworkers have reported a DCC system based upon dimerization of the highly potent antibiotic vancomycin [42]. In contrast to the systems described thus far, which involve small-molecule components equilibrating in the presence of a large biological macromolecule (termed casting) [7], the Nicoloau group examined the effect of small peptide sequences on the dimerization of the much larger vancomycin molecule (molding). The antibiotic activity of vancomycin arises from inhibition of peptidoglycan biosynthesis, which occurs though binding of the terminal Lys-D-Ala-D-Ala piece of the peptidoglycan precursor. The role of vanco-mycin dimerization in this binding event has received significant study, with the dimer having greater affinity for its target than the monomer. Two different types of DCL were designed around vancomycin analogs being derivatized at the carbohydrate moiety with thiols of varying chain length, for disulfide bond formation, as well as alkenes with the aim of establishing crossmetathesis as a synthetic method for biological DCC (Scheme 2.9). The vancomycin structure was also varied at the N-terminus with six dif-ferent amino acid residues to provide further diversity.

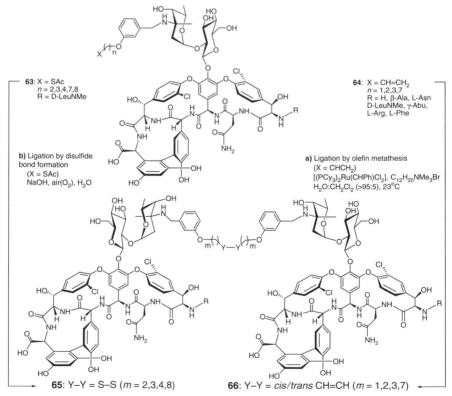

Scheme 2.9 DCLs of vancomycin dimers using crossmetathesis and disulfide bond formation.

Initial work used individual compounds **63** and **64** and examined their rate of dimerization in the presence of Ac-D-Ala-D-Ala and Ac_2-L-Lys-D-Ala-D-Ala under appropriate disulfide and crossmetathesis conditions. As expected, a strong acceleration was observed in the presence of the vancomycin target Ac_2-L-Lys-D-Ala-D-Ala. To synthesize a DCL, three of the leucine vancomycin alkene analogues were equilibrated via crossmetathesis in the presence and absence of the peptide targets. The blank library was analyzed by direct MS, which showed a mixture of intensities for all six dimers corresponding to the expected statistical weighting. Presumably the ionization profile of each dimer is very similar, owing to relatively small changes in the overall vancomycin skeleton, allowing a direct comparison of their peak sizes. In the presence of Ac_2-L-Lys-D-Ala-D-Ala the library composition was biased toward dimers having shorter tether lengths. Resynthesis and biological assay against vancomycin-resistant bacteria revealed a correlation between tether length and activity, with the shorter tether lengths having the strongest activity.

The approach was then widened to include eight vancomycin components, which would dimerise to afford 30 distinct dimers. MS analysis established that $6\text{-}(LeuNCH_3)C_2\text{-}(LeuNCH_3)C_2$ underwent the highest amplification. A wide selection of dimers was resynthesized and assayed for antibacterial activity, with excellent correlation being observed between the trends of amplification and activity.

Use of crossmetathesis as a DCL-forming reaction is a notable innovation. The reaction was first investigated in the "receptor-assisted combinatorial synthesis" context by Benner, who investigated the scope and functional group tolerance of the reaction with mixtures of alkenes [43]. Poulsen and Bornaghi have used crossmetathesis to prepare a pre-equilibrated library of dimeric CA inhibitors [44]. An immobilized version of Grubbs first-generation catalyst was used to generate DCLs, and detailed MS analysis was performed to verify reversibility. Freezing the reaction by simple filtration afforded a static library of alkenes that could be screened for activity by assay/deconvolution procedures. Applications of metathesis to adaptive protein-directed DCLs have yet to appear. Given advances in aqueous metathesis chemistry, developments in this area may be expected in the near future [45].

Greaney and coworkers have introduced the conjugate addition of thiols to Michael acceptors as an effective adaptive DCL strategy [46,47]. The reaction is well suited for biological DCL synthesis, taking place in water with no requirement for external reagents. As with disulfide bond formation, the reaction is subject to simple and effective pH control. Under mildly basic conditions, the thiolate anion adds rapidly to Michael acceptors under equilibrium conditions. Acidification effectively switches the reaction off, causing it to become extremely slow and irreversible. The reaction is well described in biological systems; the ubiquitous tripeptide glutathione (GSH), for example, is well known to conjugate Michael acceptors in the cell as part of excretory metabolism [48].

The target selected for thiol conjugate addition DCC was glutathione S-transferase (GST) from the helminth worm *Schistosoma japonicum*. The GSTs catalyze the conjugation of GSH to a wide variety of electrophilic substrates in the cell, and are an emerging drug target in cancer therapy and parasitic diseases such as malaria and schistosomiasis. An initial DCL was constructed using GSH (**67**), three tripeptide analogues that differ at the γ-glutamyl residue **68–70**, and the Michael acceptor ethacrynic acid (EA, **71**), a known inhibitor of the GST class (Scheme 2.10). Equilibration over 1 hour afforded the four expected EA adducts **72–75**, identified by LC-MS. Addition of SjGST dramatically amplified the native substrate GS–EA (**72**) from 35% of total adduct concentration to 92% under conditions where the protein was present from the start, and when it was added postequilibration. Subsequent binding assays established an order of magnitude difference between amplified **72** ($IC_{50} = 0.32$ μM) and the nonamplified component **75** (88 μM).

The thiol conjugate addition DCC was then extended to a larger library consisting of a single thiol GSH and 14 EA analogues. Whereas the initial DCL probed the GSH-binding site, a region of the SjGST protein known to be selective for the native ligand GSH and thus expected to be highly discriminatory in a DCC experiment, the DCL in Scheme 2.11 addresses the hydrophobic substrate site of SjGST that is far more lax in its selectivity. The application of DCL to such targets, where the answer is not known in advance, the resulting DCL was constructed as before, and SjGST was found to amplify the three compounds **77a**, **77m**, and **77n** at the expense of the lysine derivative **77f**. Biological assay established the amplified components to be an order of magnitude more potent ($IC_{50} = 0.61$ μM for **77a**) than the lysine component **77f** ($IC_{50} = 8.20$ μM), which is removed from the equilibrium composition of the DCL as a result.

Sasaki and coworkers have examined reversible metal coordination as a mechanism for DCL generation in the presence of lectin biomolecules [49,50]. The use of metal ions in reversible processes is canonical to supramolecular chemistry, and has been explicitly demonstrated for double-level orthogonal DCLs by Lehn and Eliseev [51]. Sasaki's system is designed around octahedral Fe(II) bipyridine complexes. The bipyridine-modified *N*-acetylgalactosamine (bipy-GalNAc) (**78**) was found to trimerize in the presence of Fe(II) to afford a 3:1 mixture of the *fac* (**79**) and *mer* (**80**) diastereoisomers, each as a racemic mixture (Λ+Δ) (Scheme 2.12).

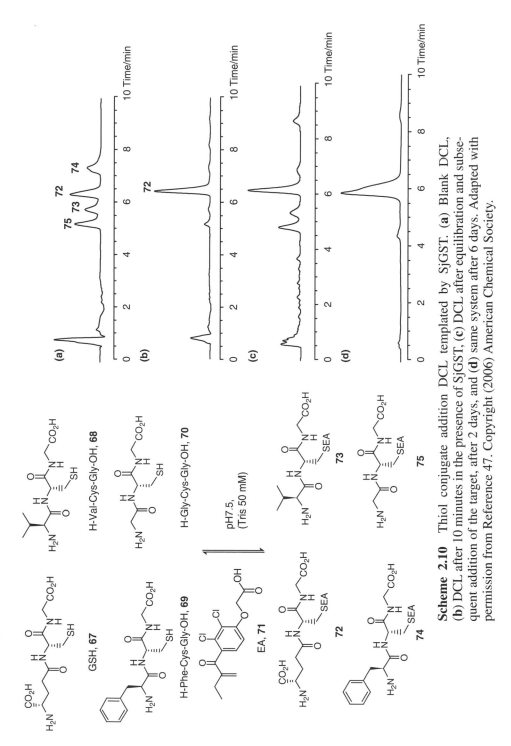

Scheme 2.10 Thiol conjugate addition DCL templated by SjGST. (**a**) Blank DCL, (**b**) DCL after 10 minutes in the presence of SjGST, (**c**) DCL after equilibration and subsequent addition of the target, after 2 days, and (**d**) same system after 6 days. Adapted with permission from Reference 47. Copyright (2006) American Chemical Society.

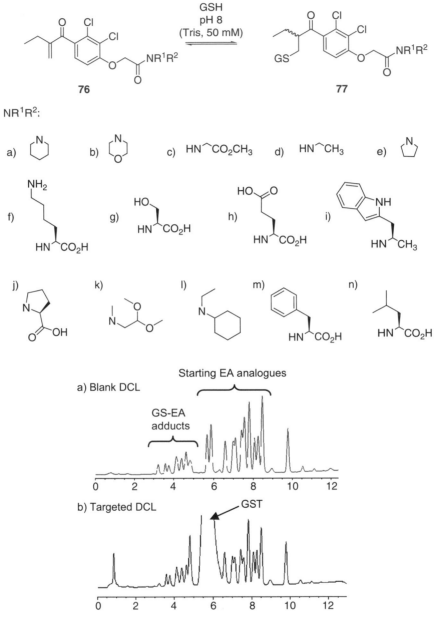

Scheme 2.11   Thiol conjugate addition DCL templated by SjGST. Trace a) shows the blank DCL assembled in the absence of enzyme. Trace b) shows the templated DCL. Adapted with etc.

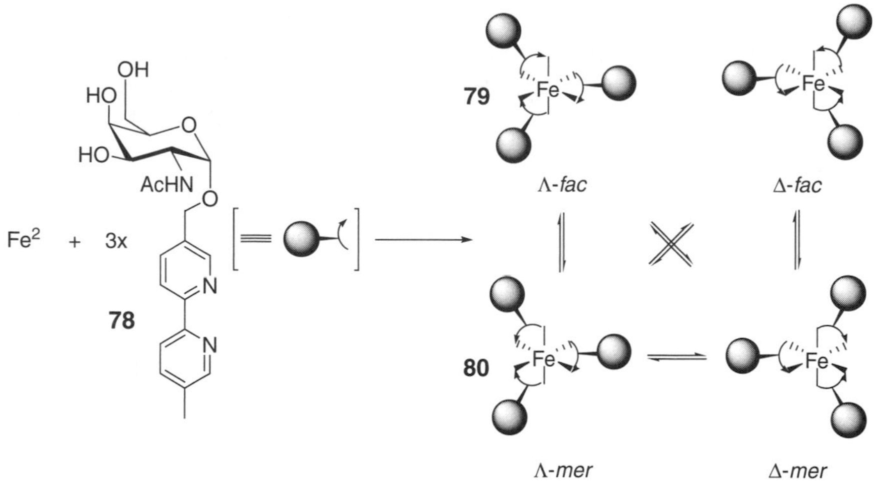

Scheme 2.12   DCL of Fe(II)(bipy-GalNAc)$_3$ complexes.

The metal complexes are in dynamic equilibrium at room temperature, and could be analyzed by chiral HPLC such that all four stereoisomers were identified. The biological template for the DCL was the GalNAc-specific plant lectin *Vicia villosa* B$_4$, known to bind the cancer-specific antigen T$_N$ through recognition of α-GalNAc-serine or threonine residues. For the system at hand, the binding of Fe(II)(bipy-GalNAc)$_3$ to *V. villosa* B$_4$ lectin was observed to be ca. seven times greater than the individual bipy-GalNAc ligand, due to the clustering effect of the GalNAc residues on the metal template. When the DCL was prepared in the presence of the lectin, the Λ-*mer* stereoisomer was slowly amplified up to 85% of the total isomer ratio over 32 hours. This amplification was correlated with binding affinity, with the relative binding constants for Δ-*fac*, Λ-*fac*, Δ-*mer*, and Λ-*mer* measured as 5.4, 1.0, 1.2, and 18, respectively. Evidence for the specific binding of Fe(II)(bipy-GalNAc)$_3$ GalNAc residues to the lectin template being responsible for amplification was strengthened by a control experiment, which added a large excess of GalNAc to the DCL. The excess free sugar inhibits any specific lectin-complex interactions, and affords the same statistical distribution of **79** and **80** observed in the blank library.

2.6. Conclusions

The field of protein-directed DCC has grown significantly since the first studies reported by Lehn and Huc in 1997 [1]. The scope of the field has now been widely studied in terms of synthetic chemistry, DCL design and analysis, and protein target. The $C=N$ double-bond-forming reactions that were first used to exemplify the DCC idea have been demonstrated to work with success, provided that the imine linkage is not too dissimilar in binding properties to the amine group produced in the final analysis. Meanwhile, enzymatic reactions and sulfur chemistry have emerged as viable alternatives for the production of genuinely adaptive DCLs. In terms of DCL design, the fact that a protein-directed DCC experiment brings two (or more) molecules together in a protein's active site makes it particularly well suited to "bi-substrate" protein targets. The bi-substrate approach, whereby an active site featuring two linked binding regions is interrogated with small molecular weight fragment molecules, is well described in the conventional medicinal chemistry literature and has provided a useful framework for applying DCC in the medicinal chemistry context.

DCL stoichiometry is integral to the design of a productive system and many variations have been tried. Two limiting systems can be identified. Simple $n \times n$ systems maximize the combinatorial dimension of DCC, but limit the extent of amplification that can be achieved for a given experiment. No matter how strong the best binder, there will always be significant amounts of weaker binding DCL "products" present at equilibrium with the protein target. At the other end of the spectrum, $n \times 1$ systems use a biologically active scaffold that can be derivatized at a single position with n components. Here, the combinatorial aspect is diminished, as the constitutional size of the DCL is no bigger than $n + 1$. Amplifications, however, can be very strong—it is possible in principle to amplify the entire library to a single component. If multivalency can be incorporated into the "1" component in this latter case, say a scaffold containing two aldehydes, then the combinatorial increase in library size can be harnessed while maintaining the strongly amplifying adaptive behaviour of the DCL.

The growth of DCC has likewise identified limitations that will challenge the development of DCC in the coming years. While the requirement for a stoichiometric amount of protein target appears inherent to the DCC concept, several systems have been described that point to creative solutions to this constraint. For example, the synthesis of DCL components can be separated from the molecular recognition event, or secondary irreversible chemical processes may be coupled to the primary DCL equilibration event. In general, however, the requirement for stoichiometric quantities of the template has led to early-phase DCC systems that use venerable, structurally characterized proteins to direct the evolution of the DCL. The protein targets are often very well understood in terms of protein–ligand interactions, with excellent medicinal chemistry SAR data available to bias the composition of the DCLs. In the future, the field must evolve to target new proteins that offer greater challenges in terms of uncovering new protein-binding molecules and studying novel protein–ligand interaction systems.

References

1. Huc, I.; Lehn, J.-M. Virtual combinatorial libraries: Dynamic generation of molecular and supramolecular diversity by self-assembly. *Proc. Natl. Acad. Sci. U.S.A.* **1997**, *94*, 2106–2110.

2. Corbett, P. T.; Leclaire, J.; Vial, L.; West, K. R.; Wietor, J.-L.; Sanders, J. K. M.; Otto, S. Dynamic combinatorial chemistry. *Chem. Rev.* **2006**, *106*, 3652–3711.

3. Otto, S.; Furlan, R. L. E.; Sanders, J. K. M. Recent developments in dynamic combinatorial chemistry. *Curr. Opin. Chem. Biol.* **2002**, *6*, 321–327.

4. Ramström, O.; Lehn, J.-M. Drug discovery by dynamic combinatorial libraries. *Nat. Rev. Drug Discov.* **2002**, *1*, 26–36.

5. Otto, S.; Furlan, R. L. E.; Sanders, J. K. M. Dynamic combinatorial chemistry. *Drug Discov. Today* **2002**, *7*, 117–125.

6. Karan, C.; Miller, B. L. Dynamic diversity in drug discovery: Putting small-molecule evolution to work. *Drug Discov. Today* **2000**, *5*, 67–75.

7. Lehn, J.-M. Dynamic combinatorial chemistry and virtual combinatorial libraries. *Chem. Eur. J.* **1999**, *3*, 2455–2463.

8. Bunyapaiboonsri, T.; Ramström, O.; Lohmann, S.; Lehn, J.-M.; Peng, L.; Goeldner, M. Dynamic deconvolution of a pre-equilibrated dynamic combinatorial library of acetylcholinesterase inhibitors. *Chembiochem* **2001**, *2*, 438–444.

9. Cousins, G. R. L.; Poulsen, S.-A.; Sanders, J. K. M. Dynamic combinatorial libraries of pseudo-peptide hydrazone macrocycles. *Chem. Commun.* **1999**, 1575–1576.

10. Bunyapaiboonsri, T.; Ramström, H.; Ramström, O.; Haiech, J.; Lehn, J.-M. Generation of bis-cationic heterocyclic inhibitors of *Bacillus subtilis* HPr kinase/phosphatase from a ditopic dynamic combinatorial library. *J. Med. Chem.* **2003**, *46*, 5803–5811.

11. Hochgürtel, M.; Kroth, H.; Piecha, D.; Hofmann, M. W.; Nicolau, C.; Krause, S.; Schaaf, O.; Sonnenmoser, G.; Eliseev, A. V. Target-induced formation of neuraminidase inhibitors from in vitro virtual combinatorial libraries. *Proc. Natl. Acad. Sci. U.S.A.* **2002**, *99*, 3382–3387.

12. Hochgürtel, M.; Biesinger, R.; Kroth, H.; Piecha, D.; Hofmann, M. W.; Krause, S.; Schaaf, O.; Nicolau, C.; Eliseev, A. V. Ketones as building blocks for dynamic combinatorial libraries: Highly active neuraminidase inhibitors

generated via selection pressure of the biological target. *J. Med. Chem.* **2003**, *46*, 356–358.

13. Zameo, S.; Vauzeilles, B.; Beau, J.-M. Dynamic combinatorial chemistry: Lysozyme selects an aromatic motif that mimics a carbohydrate residue. *Angew. Chem. Int. Ed.* **2005**, *44*, 965–969.

14. Zameo, S.; Vauzeilles, B.; Beau, J.-M. Direct composition analysis of a dynamic library of imines in an aqueous medium. *Eur. J. Org. Chem.* **2006**, 5441–5444.

15. Valade, A.; Urban, D.; Beau, J.-M. Target-assisted selection of galactosyltransferase binders from dynamic combinatorial libraries. An unexpected solution with restricted amounts of the enzyme. *Chembiochem* **2006**, *7*, 1023–1027.

16. Valade, A.; Urban, D.; Beau, J.-M. Two galactosyltransferases' selection of different binders from the same uridine-based dynamic combinatorial library. *J. Comb. Chem.* **2007**, *9*, 1–4.

17. Congreve, M. S.; Davis, D. J.; Devine, L.; Granata, C.; O'Reilly, M.; Wyatt, P. G.; Jhoti, H. Detection of ligands from a dynamic combinatorial library by X-ray crystallography. *Angew. Chem. Int. Ed.* **2003**, *42*, 4479–4482.

18. Katz, B. A.; Finer-Moore, J.; Mortezaei, R.; Rich, D. H.; Stroud, R. M. Episelection: Novel K_i approximately nanomolar inhibitors of serine proteases selected by binding or chemistry on an enzyme surface. *Biochemistry* **1995**, *34*, 8264–8280.

19. Hioki, H.; Still, W. C. Chemical evolution: A model system that selects and amplifies a receptor for the tripeptide (D)Pro(L)Val(D)Val. *J. Org. Chem.* **1998**, *63*, 904–905.

20. Otto, S.; Furlan, R. L. E.; Sanders, J. K. M. Dynamic combinatorial libraries of macrocyclic disulfides in water. *J. Am. Chem. Soc.* **2000**, *122*, 12063–12064.

21. Ramström, O.; Lehn, J.-M. In situ generation and screening of a dynamic combinatorial carbohydrate library against concanavalin A. *Chembiochem* **2000**, *1*, 41–48.

22. Ramström, O.; Lohmann, S.; Bunyapaiboonsri, T.; Lehn, J.-M. Dynamic combinatorial carbohydrate libraries: Probing the binding site of the concanavalin A lectin. *Chem. Eur. J.* **2004**, *10*, 1711–1715.

23. Milanesi, L.; Hunter, C. A.; Sedelnikova, S. E.; Waltho, J. P. Amplification of bifunctional ligands for calmodulin from a dynamic combinatorial library. *Chem. Eur. J.* **2006**, *12*, 1081–1087.

24. Danieli, B.; Giardini, A.; Lesma, G.; Passarella, D.; Peretto, B.; Sacchetti, A.; Silvani, A.; Pratesi, G.; Zunino, F. Thiocolchicine–podophyllotoxin conjugates: Dynamic libraries based on disulfide exchange reaction. *J. Org. Chem.* **2006**, *71*, 2848–2853.

25. Erlanson, D. A.; Ballinger, M. D.; Wells, J. A. In *Fragment-based approaches in drug discovery*. In: Jahnke, W.; Erlanson, D. A. editors. Wiley-VCH, Weinheim, Tethering, 285–308.

26. Braisted, A. C.; Oslob, J. D.; Delano, W. L.; Hyde, J.; McDowell, R. S.; Waal, N.; Yu, C.; Arkin, M. R.; Raimundo, B. C. Discovery of a potent small molecule IL-2 inhibitor through fragment assembly. *J. Am. Chem. Soc.* **2003**, *125*, 3714–3715.

27. Erlanson, D. A.; Lam, J. W.; Wiesmann, C.; Luong, T. N.; Simmons, R. L.; Delano, W. L.; Choong, I. C.; Burdett, M. T.; Flanagan, W. M.; Lee, D.; Gordon, E. M.; O'Brien, T. O. In situ assembly of enzyme inhibitors using extended tethering. *Nat. Biotechnol.* **2003**, *21*, 308–314.

28. Hardy, J. A.; Lam, J.; Nguyen, J. T.; O'Brien, T.; Wells, J. A. Discovery of an allosteric site in the caspases *Proc. Nat. Acad. Sci. U.S.A.* **2004**, *101*, 12461–12466.

29. Buck, E.; Bourne, H.; Wells, J. A. Site-specific disulfide capture of agonist and antagonist peptides on the C5a receptor. *J. Biol. Chem.* **2005**, *280*, 4009–4012.

30. Buck, E.; Wells, J. A. Disulfide trapping to localize small-molecule agonists and antagonists for a G protein-coupled receptor. *Proc. Natl. Acad. Sci. U.S.A.* **2005**, *102*, 2719–2724.

31. Erlanson, D. A.; Braisted, A. C.; Raphael, D. R.; Randal, M.; Stroud, R. M.; Gordon, E. M.; Wells, J. A. Site-directed ligand discovery. *Proc. Nat. Acad. Sci. U.S.A.* **2000**, *97*, 9367–9372.

32. Swann, P. G.; Casanova, R. A.; Desai, A.; Frauenhoff, M. M.; Urbancic, M.; Slomczynska, U.; Hopfinger, A. J.; Le Breton, G. C.; Venton, D. L. Nonspecific protease-catalyzed hydrolysis/synthesis of a mixture of peptides: Product diversity and ligand amplification by a molecular trap. *Biopolymers* **1997**, *40*, 617–625.

33. Hoerter, J. M.; Otte, K. M.; Gellman, S. H.; Cui, Q.; Stahl, S. S. Discovery and mechanistic study of Al^{III}-catalyzed transamidation of tertiary amides. *J. Am. Chem. Soc.* **2008**, *130*, 647–654.

34. Eldred, S. E.; Stone, D. A.; Gellman, S. H.; Stahl, S. S. Catalytic transamidation under moderate conditions. *J. Am. Chem. Soc.* **2003**, *125*, 3422–3423.

35. Bell, C. M.; Kissounko, D. A.; Gellman, S. H.; Stahl, S. S. Catalytic metathesis of simple secondary amides. *Angew. Chem. Int. Ed.* **2007**, *46*, 761–763.

36. Lins, R. J.; Flitsch, S. L.; Turner, N. J.; Irving, E.; Brown, S. A. Enzymatic generation and in situ screening of a dynamic combinatorial library of sialic acid analogues. *Angew. Chem. Int. Ed.* **2002**, *41*, 3405–3407.

37. Lins, R. J.; Flitsch, S. L.; Turner, N. J.; Irving, E.; Brown, S. A. Generation of a dynamic combinatorial library using sialic acid aldolase and in situ screening against wheat germ agglutinin. *Tetrahedron* **2004**, *60*, 771–780.

38. Cheeseman, J. D.; Corbett, A. D.; Shu, R.; Croteau, J.; Gleason, J. L.; Kazlauskas, R. J. Amplification of screening sensitivity through selective destruction: Theory and screening of a library of carbonic anhydrase inhibitors. *J. Am. Chem. Soc.* **2002**, *124*, 5692–5701.

39. Corbett, A. D.; Cheeseman, J. D.; Kazlauskas, R. J.; Gleason, J. L. Pseudodynamic combinatorial libraries: A receptor-assisted approach for drug discovery. *Angew. Chem. Int. Ed.* **2004**, *43*, 2432–2436.

40. Eliseev, A. V.; Nelen, M. I. Use of molecular recognition to drive chemical evolution. 1. Controlling the composition of an equilibrating mixture of simple arginine receptors. *J. Am. Chem. Soc.* **1997**, *119*, 1147–1148.

41. Eliseev, A. V.; Nelen, M. I. Use of molecular recognition to drive chemical evolution. 2. Mechanisms of an automated genetic algorithm implementation. *Chem. Eur. J.* **1998**, *4*, 825–834.

42. Nicolaou, K. C.; Hughes, R.; Cho, S. K.; Winssinger, N.; Smethurst, C.; Labischinski, H.; Endermann, R. *Angew. Chem. Int. Ed.* **2000**, *39*, 3823–3828.

43. Giger, T.; Wigger, M.; Audétat, S.; Benner, S. A. Libraries for receptor-assisted combinatorial synthesis (RACS). The olefin metathesis reaction. *Synlett* **1998**, 688–691.

44. Poulsen, S.-A.; Bornaghi, L. F. Fragment-based drug discovery of carbonic anhydrase II inhibitors by dynamic combinatorial chemistry utilizing alkene cross metathesis. *Bioorg. Med. Chem.* **2006**, *14*, 3275–3284.

45. Jordan, J. P.; Grubbs, R. H. Small-molecule N-heterocyclic-carbene-containing olefin-metathesis catalysts for use in water. *Angew. Chem. Int. Ed.* **2007**, *46*, 5152–5155.

46. Shi, B.; Greaney, M. F. Reversible Michael addition of thiols as a new tool for dynamic combinatorial chemistry. *Chem. Commun.* **2005**, 886–888.

47. Shi, B.; Stevenson, R.; Campopiano, D. J.; Greaney, M. F. Discovery of glutathione S-transferase inhibitors using dynamic combinatorial chemistry. *J. Am. Chem. Soc.* **2006**, *128*, 8459–8467.

48. van Bladeren, P. J. Glutathione conjugation as a bioactivation reaction. *Chem. Biol. Int.* **2000**, *129*, 61–76.

49. Sakai, S.; Shigemasa, Y.; Sasaki, T. A self-adjusting carbohydrate ligand for GalNAc specific lectins. *Tetrahedron Lett.* **1997**, *38*, 8145–8148.

50. Sakai, S.; Shigemasa, Y.; Sasaki, T. Iron(II)-assisted assembly of trivalent GalNAc clusters and their interactions with GalNAc-specific lectins. *Bull. Chem. Soc. Jpn.* **1999**, *72*, 1313–1319.

51. Goral, V.; Nelen, M. I.; Eliseev, A. V.; Lehn, J.-M. Double-level "orthogonal" dynamic combinatorial libraries on transition metal template. *Proc. Natl. Acad. Sci. U.S.A.* **2001**, *98*, 1347–1352.

Chapter 3

Nucleic Acid-Targeted Dynamic Combinatorial Chemistry

Peter C. Gareiss and Benjamin L. Miller

3.1. Introduction

Nucleic acids are arguably life's most fundamental building blocks. From the obvious role of DNA in biological information storage, to the many regulatory roles of RNA (with more being discovered on an almost daily basis), nucleic acids present a myriad of opportunities for using small molecules to influence the course of disease, or study essential biological functions. The modular nature of DNA structure made it an attractive target for some of the earliest dynamic combinatorial chemistry (DCC) experiments. Subsequent efforts by a number of research groups have expanded the range of nucleic acid targets successfully addressed by DCC beyond canonical B-form DNA to include a broad diversity of structures. This chapter will survey DCC systems targeting the development of ligands for DNA and RNA, to modify DNA and RNA, and to develop novel nucleic-acid-based polymers and nanostructures.

3.2. DCC Targeting DNA

The "early history" of the DCC field is coincident with the publication of a series of landmark papers from the Dervan laboratory, describing a "code" for the design and synthesis of polypyrrole–polyimidazole compounds

Dynamic Combinatorial Chemistry, edited by Benjamin L. Miller
Copyright © 2010 John Wiley & Sons, Inc.

capable of recognizing any arbitrary duplex DNA sequence with exquisite selectivity [1]. The Dervan group's work was the spectacular culmination of decades of effort by a number of laboratories in academia and industry to develop small molecules[1] capable of selectively binding DNA sequences, both as a fundamental challenge in molecular recognition, and as a chemical means to alter and control gene expression. To date, the wide variety of DNA sequences targeted by small molecules includes promoter regions, transcriptional elements such as the TATA box, and structural elements such as G-quadruplexes.

The design of synthetic DNA ligands was in turn driven and guided by the discovery of many natural products that selectively bind DNA sequences [2]. Two common classes of DNA-binding natural products are intercalators (such as bisintercalators echinomycin and triostin A) and polyamides (such as distamycin A and netropsin) (Fig. 3.1) [3]. Aspects of these natural products have aided in the development of novel synthetic DNA ligands. The set of rules developed by the Dervan lab for the tunable and programmable synthesis of efficient and specific DNA-binding polyamides allows selectively targeting each possible Watson–Crick base pair by combining pyrrole-, imidazole-, and hydroxypyrrole-based amino acids [4,5]. These

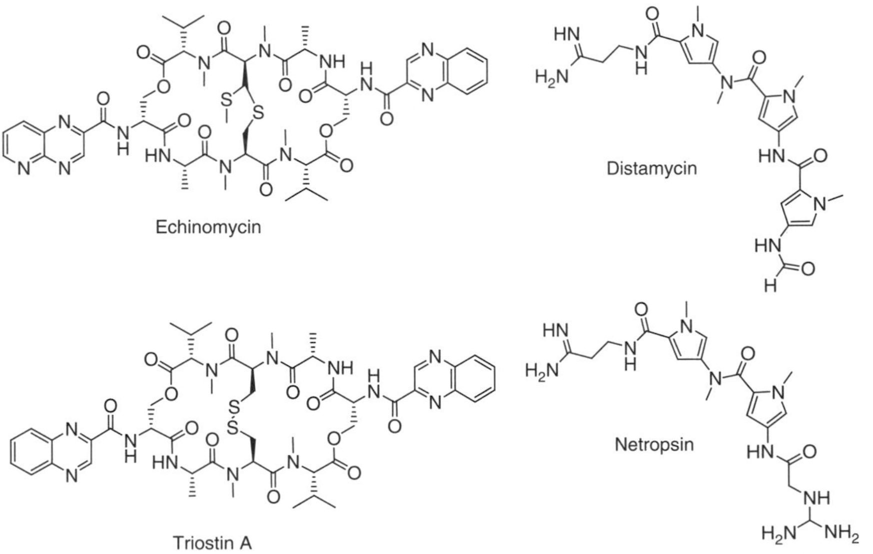

Echinomycin

Distamycin

Triostin A

Netropsin

Figure 3.1 DNA-binding natural products that have inspired designed ligands for oligonucleotides.

[1] Throughout this chapter, we use the term "small molecule" to also encompass peptides and peptide/small-molecule hybrids.

polyamides are cell and nuclear permeable, and have proven effective at repressing, regulating, and activating native gene expression.

The first dynamic combinatorial efforts targeting DNA were reported in 1997 by Miller and coworkers [6]. Conceptually, the goal of these initial studies was to use reversible metal ion coordination to generate dynamic libraries of coordination complexes. Subsequent incubation of the equilibrating library with a resin-immobilized DNA target was anticipated to allow for selection and amplification of the highest affinity ligand. In practice, initial experiments employed reversible salicylaldimine–metal complexation to develop ligands for cellulose-immobilized oligo d(A · T) DNA (Fig. 3.2).

Using this approach, a dynamic combinatorial library (DCL) of rapidly equilibrating metal complexes was formed by incubating a series of salicylaldimines with the transition metal salt $ZnCl_2$ in aqueous solution (Fig. 3.3). Divalent zinc was chosen as the transition metal for its known tetravalent coordination geometry with salicylaldimines [7] and its compatibility with DNA. (In retrospect, this was a somewhat naïve view of both

Figure 3.2 Schematic of Miller's metal complexation/affinity reagent dynamic combinatorial strategy. A series of monomers is allowed to rapidly equilibrate by metal complexation. Addition of an immobilized receptor drives the selection of metal complexes with affinity for the immobilized receptor.

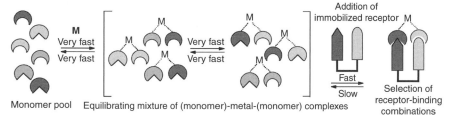

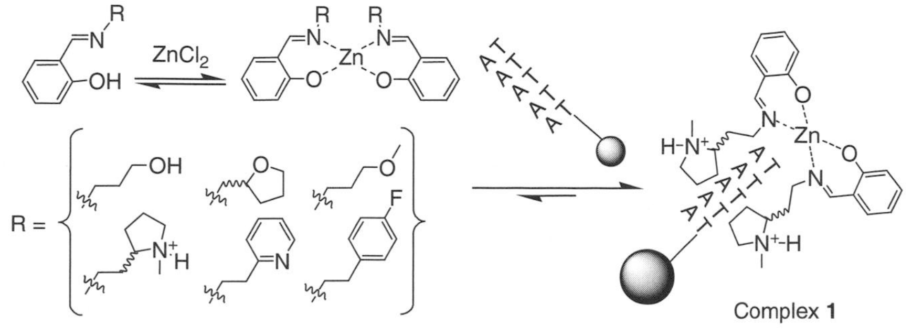

Figure 3.3 Dynamic bis-salicylaldimine Zn^{2+} complexation identifies an oligo d(A · T) DNA ligand. Right: The selected N-methyl 2-aminoethyl pyrrolidine based bis-salicylaldimine complex **1** (hypothetical), which binds to oligo d(A · T) DNA.

the stability of salicylaldimines and the coordination chemistry of zinc; *vide infra.*) First, salicylaldimine complexation with Zn^{2+} was confirmed by NMR. Next, a dynamic library of salicylaldimines was incubated with the oligo $d(A \cdot T)$ DNA cellulose resin both in the presence and absence of Zn^{2+}.

Contrary to many DCC experiments, which undergo analysis based on direct observation of amplification factors corresponding to selected compounds, analysis of this affinity reagent DCC approach was based on depletion factors, since the selected compounds are bound by the affinity reagent and removed from the DCL. In this approach, the library analysis consisted of first eluting the DCL solution from the affinity resin. The resulting library mixture was then hydrolyzed and derivatized with naphthoyl chloride to stop equilibration, and the ratios of compounds retained on the affinity column in the presence and absence of $ZnCl_2$ were determined by HPLC analysis. The process revealed that two of the six possible amines (and, presumed by extension, their Zn–salicylaldimine complexes) were bound by the target oligo $d(A \cdot T)$ DNA resin: *N*-methyl 2-aminoethyl pyrrolidine and 2-aminomethyl furan. Further deconvolution and UV titration experiments led to the conclusion that the homodimeric *N*-methyl 2-aminoethyl pyrrolidine based bis-salicylaldimine complex **1** was the best binder and had a dissociation constant to the oligo $d(A \cdot T)$ DNA target of 1.1 μM. As the only member of the library bearing a positive charge at neutral pH, this was an expected result; nevertheless, it provided substantial proof-of-concept validation for the approach.

Further consideration of this experimental system led Miller and coworkers to the conclusion that the library was substantially more complex than that initially hypothesized. While it was assumed coordination to zinc would eliminate (or at least greatly reduce) salicylaldimine hydrolysis, subsequent NMR experiments on the putative complex **1** revealed this to be an incorrect assumption. This observation led to the hypothesis that one should be able to observe similar selectivities via a "multistage" DCL selection, in which amines, salicylaldehydes, and zinc would be mixed, and allowed to undergo selection in the presence of DNA [8] (Fig. 3.4).

As in the initial report, this multistage DCL was screened against oligo $d(A \cdot T)$ DNA resin. Consistent with the "multistage" hypothesis, *N*-methyl 2-aminoethyl pyrrolidine was again selected as the highest affinity library member. This approach proved that multistage equilibration was applicable in dynamic combinatorial experiments utilizing immobilized targets, and paved the way for future expansion of dynamic processes [9,10]. However, while these studies served as useful "proof-of-concept" demonstrations of DNA-targeted DCC selections, they are also illustrative of several important design issues that must be carefully considered for any DCL. In addition to ensuring that the exchange reaction is incapable of modifying the target

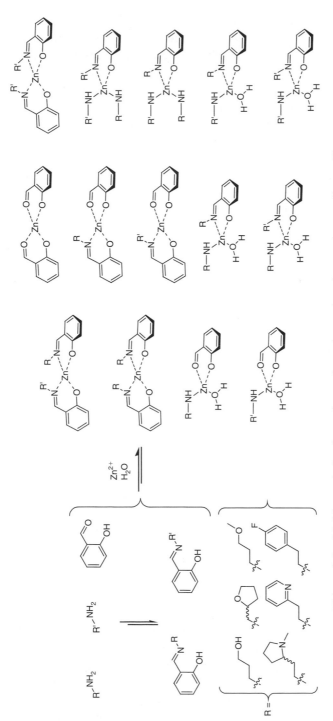

Figure 3.4 Multistage imine formation and metal complexation produces a diverse library of amine–, salicylaldehyde–, and salicylaldimine–zinc complexes.

(and carried out under conditions hospitable to the target), it is equally important to ensure that the exchange reaction (or reactions) allows for unambiguous identification of the selected compound following amplification. Here, this was made difficult by the reversibility of imine formation (used to good effect by several other groups in DCC experiments) coupled with the rapid mutability of zinc complexes.

Subsequent DNA-directed efforts by the Miller group and others have focused on exchange chemistry much more amenable to unambiguous product analysis. In particular, disulfide exchange has proven to be an exceptionally robust and simple method for library equilibration. In 2004, Balasubramanian and coworkers reported a disulfide-based DCL targeting G-quadruplex DNA [10]. The G-quadruplex DNA motif is formed by stacking interactions between tetrameric G platforms, and this structure is involved in a variety of biological processes. Importantly, G-quadruplex DNA plays a role in chromosomal alignment and teleomere formation, processes that are involved in gene regulation and disease [11]. Thus, G-quadruplex DNA has become an attractive target for a number of groups [12]. In the 2004 Balasubramanian group DCL approach, it was hypothesized that high affinity G-quadruplex ligands would likely be formed by joining two groups known to have affinity for the G-quadruplex structure (Fig. 3.5): a G-tetrad hydrophobic interacting moiety (**A**), and an electrostatic and base interacting peptidic moiety (**P**). To this end, an acridone unit (**A**) designed to hydrophobically interact with the G-tetrad was allowed to dynamically equilibrate via disulfide exchange with a quadruplex recognizing tetrapeptide (**P**) of sequence Phe-Arg-His-Arg in the presence of glutathione (**G**). Glutathione was selected as a library building block primarily to mediate disulfide exchange between **A** and **P**. Equilibration allowed a library of six possible disulfide members to be formed. The DCL was equilibrated in the presence and absence of a biotinylated human telomeric quadruplex sequence [5′-biotin-d(GTTAGG)$_5$]. Biotinylated DNA target was chosen to aid in the identification of selected DNA binding library members by a streptavidin–biotin affinity reagent approach. Briefly, the DNA–ligand complexes were captured by streptavidin beads, and the DNA was then heat denatured, freeing the bound, dynamically selected ligands identified by HPLC analysis.

Using this screening approach, a fourfold amplification of the designed library member **A–P** was observed. Interestingly, a fivefold amplification of the peptidic dimer **P–P** was also observed. Amplification factors correlated well with dissociation constants (K_D) subsequently measured by surface plasmon resonance (SPR): **A–P** was found to have a K_D of 30 μM, while the more strongly amplified **P–P** was found to have a K_D of 22.5 μM.

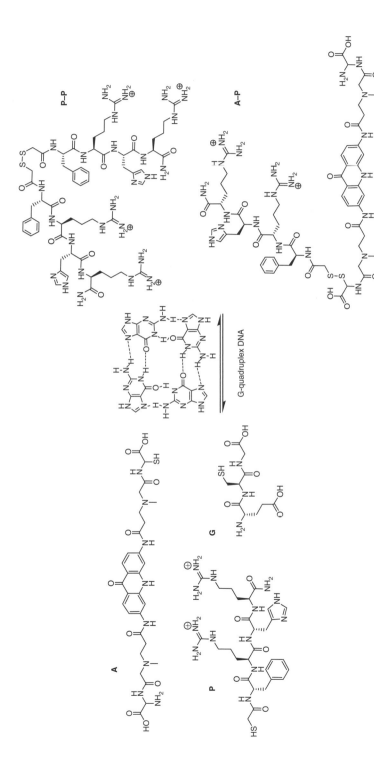

Figure 3.5 Disulfide-based dynamic combinatorial library selects for peptidic homodisulfide **P–P** and acridone–peptide heterodisulfide **A–P** G-quadruplex DNA ligands.

This relatively small DCL yielded the first peptidic ligand for G-quadruplex DNA, providing a lead compound for the further development of novel G-quadruplex ligands with possible therapeutic utility.

Revisiting this area in 2005 [13], Balasubramanian and coworkers reported a six-member DCL based on disulfide exchange between thiolated N-methylpyrrole oligomers. The N-methylpyrrole oligomer library building block design (**P1–P3**, Fig. 3.6) was greatly influenced by the natural products distamycin and netropsin, in addition to the strong and selective N-methylpyrrole and N-methylimidazole polyamide DNA ligands developed by the Dervan group [1]. The library was screened against the telomeric G-quadruplex DNA 5′-biotin-d(GTTAGG)$_5$ and an 11-*mer* A/T-rich DNA duplex sequence from the promoter region of *ckit*, a human oncogene: 5′-biotin-d(CTTTTATTTTG), hybridized with 5′-biotin-d(GAAAATAAAAC). The screening procedure, amplification process, and streptavadin-based identification methodology employed were identical to their 2004 report [13].

Screening of the library against these sequences resulted in amplification of DNA-binding ligands in both cases. Curiously, however, these amplifications were only significant for the experiment employing double-stranded DNA as target: 2.2-, 2.95-, and 6.6-fold for compounds **P2–P2**, **P3–P3**, and **P2–P3**, respectively. Screening against the G-quadruplex DNA resulted in a modest 0.3- and 0.4-fold amplification of **P2–P3** and **P3–P3**, respectively. The authors indicated that these results were not entirely surprising. The increased amplification observed when screening against double-stranded

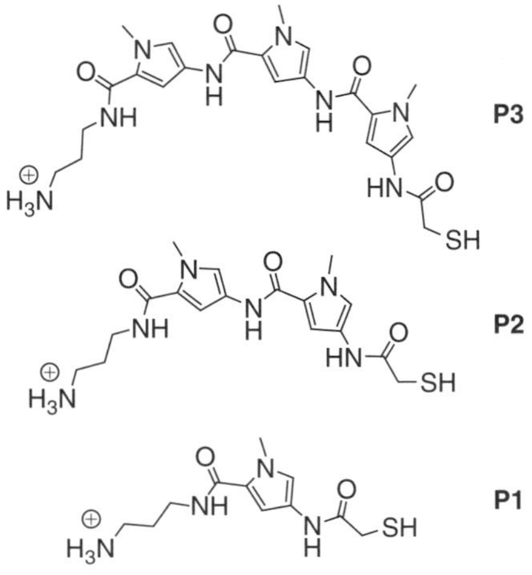

Figure 3.6 N-Methylpyrrole DCL building blocks (Balasubramanian and coworkers).

DNA over G-quadruplex DNA is consistent with the propensity of polyamides to bind to double-stranded DNA helical grooves. Although no measurements of binding affinities were performed, thermal denaturation studies showed that the stability of DNA–ligand complexes were well correlated with the observed amplification factors. An important, potentially generally applicable experimental observation in this study was that addition of competing, non-target-binding thiols (i.e., glutathione in this case) may be used to diminish self (non-target-driven) selection in disulfide libraries, and allow for amplification of the best binding disulfide species. One can speculate whether similar "equilibration enhancing" library members can be used in libraries employing reversible bond-forming reactions other than disulfide.

An alternative approach to the identification of G-quadruplex ligands has been described by Nielsen and Ulven [14]. In this study, aromatic scaffolds bearing one, two, or three disulfide-bearing arms were allowed to equilibrate three positively charged side-chain disulfides (Fig. 3.7). Extraction of G-quadruplex-binding DCL members using a procedure analogous to that employed by the Balasubramanian group allowed identification of compound **2**; a control experiment employing random-coil DNA did not select any library member. Compound **2** was subsequently shown to stabilize the melting temperature of the G-quadruplex structure by 12.8°C, a strong evidence for its binding ability.

In 2006, McNaughton and Miller disclosed a novel DCC approach termed resin-bound DCC (RBDCC) and applied its first demonstration to the development of DNA-binding compounds [15] (Fig. 3.8). Phase separations have proven useful in many DCC applications. We have already mentioned immobilization of the target in the context of DNA; other early examples include Eliseev's use of resin-immobilized guanidinium groups [9], and Still's selection of peptide-binding receptors with resin-immobilized tripeptides [16]. More recently, extraction of DCC products into a different solution [17,18] or gel phase [19] as part of the selection process has also been reported. RBDCC represents the first example of "phase-tagging" elements of a DCL. In the RBDCC approach, library constituents are immobilized on solid support, and allowed to equilibrate with an identical set of library constituents in solution via reversible bond formation. Screening in the RBDCC approach utilizes a labeled target. This separation based on phase allows any immobilized library component that binds the target to be easily visualized, physically removed from the remainder of the library, and identified. This reduces the often complex deconvolution of DCC selections to a simple matter of identifying a compound (or compounds) on a resin bead. In this respect, RBDCC overcomes the

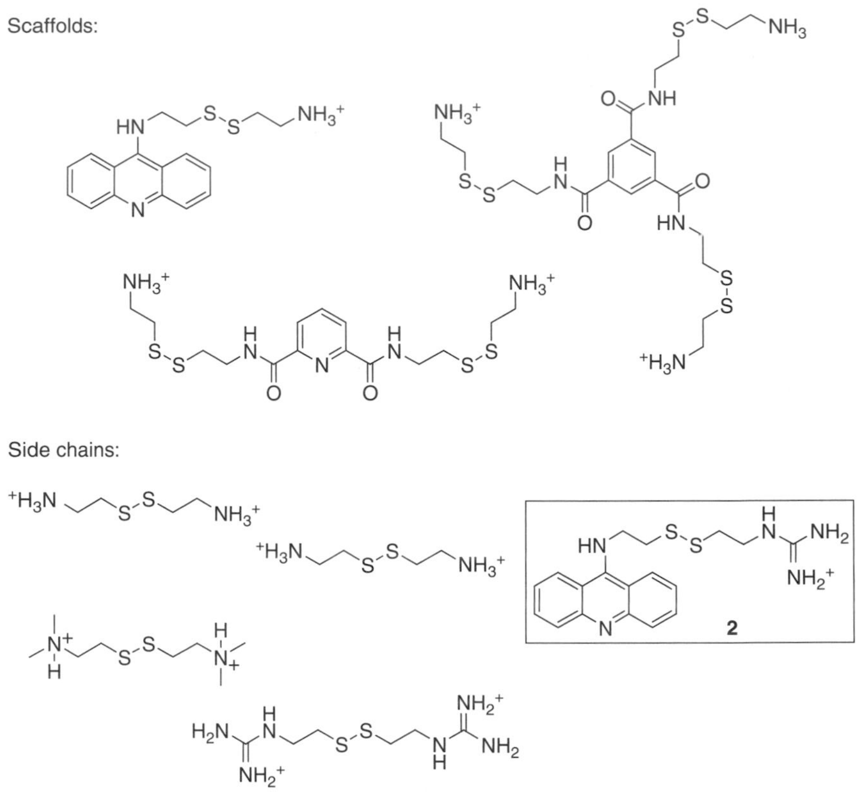

Scaffolds:

Side chains:

Figure 3.7 Scaffold-centered DCL targeting G-quadruplex DNA (Nielsen and Ulven).

common analytical hurdles associated with solution-phase DCC, including chromatographic and mass overlap.

The design of this initial resin-bound DCL (RBDCL) was influenced by the octadepsipeptide bis-intercalating DNA-binding natural products echinomycin and triostin A (Fig. 3.9). Nine thiol-based quinoline-containing tripeptide library building blocks were synthesized. These tripeptides incorporated cysteine at the first position, followed by either Gln, Ser, or His at the second and third positions, and terminated by an amide-linked 2-ethyl-quinoline (Quin) [20]. Upon disulfide exchange of the nine building blocks, a 54-member DCL was formed. In the RBDCC screening process, each of the individual batches of resin bearing one of the nine library monomeric building blocks in separate vessels were incubated with all of the nine possible solution-phase monomers and fluorescently labeled DNA target sequences. Theoretically, if only one set of resin (A) fluoresces, it

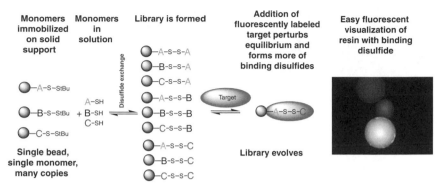

Figure 3.8   Resin-bound dynamic combinatorial chemistry. Left: Dynamic combinatorial building blocks immobilized on solid phase resin and in solution are allowed to equilibrate by reversible bond exchange to form a resin-bound dynamic combinatorial library. Center: The library is screened against a fluorescently labeled target, and dynamic selection occurs. Right: The selected library members binding to the labeled target are easily visualized, spatially segregated, and identified.

can be concluded that the selected ligand is the homodisulfide compound (A–A) from that resin. However, if two sets of resin (A and B) show fluorescence, it can be concluded that the selected ligand is either the homodisulfide (A–A or B–B) or the heterodisulfide (A–B).

The 54-member RBDCL was screened against two duplex DNA strands: 5′-TCTAGACGTC-3′ (Sequence 1) and 5′-CCATGATATC-3′ (Sequence 2). These sequences were selected because of their well-characterized history as targets for bis-intercalators. Sequence 1 is preferentially bound by triostin A, and Sequence 2 is preferentially bound by the synthetic ligand TANDEM [21,22]. First, to demonstrate the inherent challenges associated with traditional DCC, an analogous solution-phase library was prepared and screened. As seen in Fig. 3.10, superimposed HPLC traces of the solution-phase library clearly show that a product is amplified upon addition of Sequence 2, while addition of Sequence 1 produces no obvious amplification. It is equally obvious, however, that chromatographic overlap would make identification of the amplified compound challenging, at best.

Next, the RBDCL was screened against the TAMRA-labeled DNA sequences. As seen in Fig. 3.11, only one pool of resin showed significant fluorescence. This pool contained the monomer Cys-Ser-Ser-Quin, and as such the homodisulfide (Cys-Ser-Ser-Quin)$_2$ was selected as the best binder. Equilibrium dialysis experiments confirmed that (Cys-Ser-Ser-Quin)$_2$ bound the target DNA Sequence 2 with a dissociation constant of 2.8 μM. While it is certainly true that identification of amplified compounds from large solution-phase DCLs is possible, given sufficiently

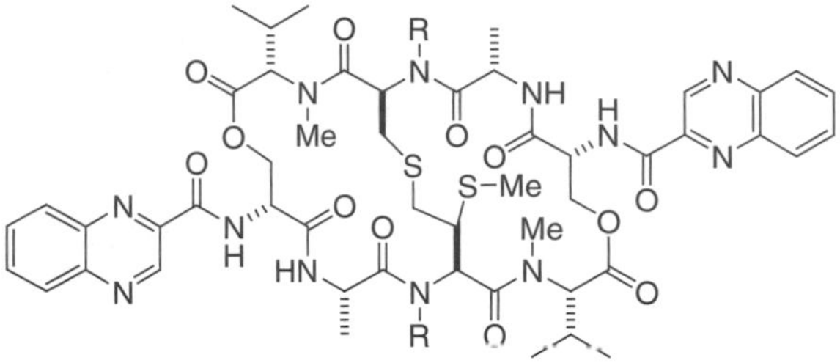

Echinomycin

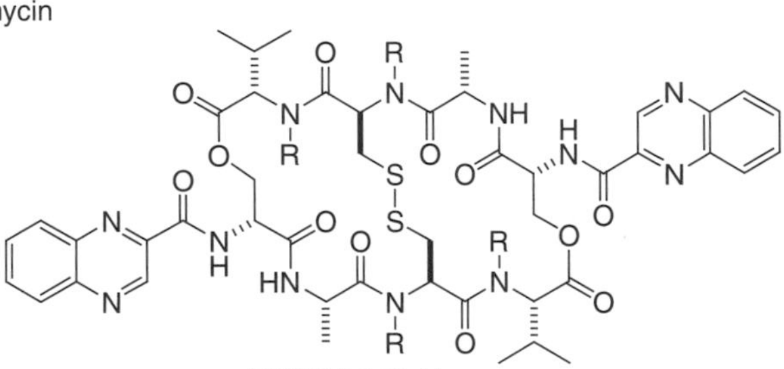

TANDEM, R=H
Triostin A, R=CH₃

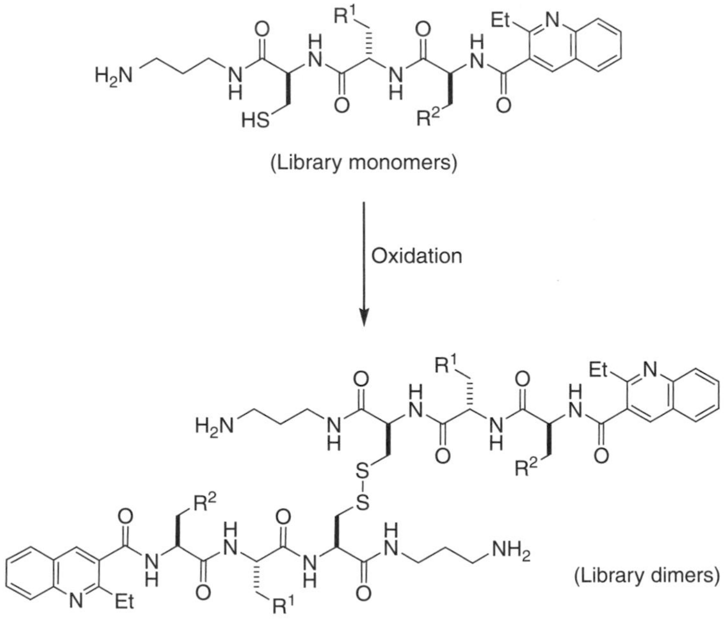

(Library monomers)

Oxidation

(Library dimers)

Figure 3.9 The design of McNaughton and Miller's quinoline-capped tripeptide dynamic combinatorial library.

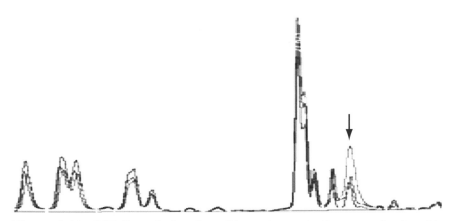

Figure 3.10 HPLC analysis of a 54-member quinoline-capped tripeptide disulfide based dynamic combinatorial library. Black: In the absence of target. Dark gray: In the presence of sequence 1. Light gray: In the presence of Sequence 2. The arrow highlights a library member selectively amplified upon addition of Sequence 2. One would anticipate that identification of the selected compound using "standard" analytical methods would be difficult.

talented analytical chemists and strong amplification, RBDCC dramatically simplifies this process. Although it is not yet clear if the RBDCC method is extendable to nondimeric libraries (a potential problem noted by others) [23], successful application to large (>11,000-compound) libraries has been achieved (discussed later in the chapter).

DCC has proven to be an effective system to develop small-molecule ligands for DNA. The salicylaldimine–metal–DNA complexation reported by Miller and coworkes provided the initial boost to target DNA with DCC, and had further applications targeting RNA as described in the following sections. Balasubramanian's natural product polyamide DCL, along with the synthetic acridone peptide DCL, were both effective in developing DNA ligands and emphasized that buffering thiols (i.e., glutathione) are efficient additives to enhance selected library members in disulfide-based DCLs. "Scaffold-centered" libraries such as those reported by Nielsen and Ulven have shown the value of screening libraries with scaffolds bearing different numbers of attachments. Finally, the novel approach of RBDCC has broadened the scope of library size and provided an efficient and easy way to deconvolute best binding library members. We can anticipate that DCC strategies will continue to play an important part in the identification of DNA-binding compounds, and particularly for unusual secondary and tertiary structures. However, as discussed in the next section, DCC approaches are becoming increasingly useful in addressing another type of nucleic acid: RNA.

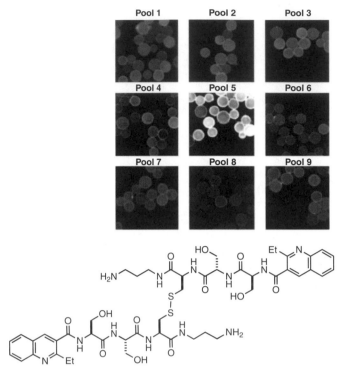

Figure 3.11 Left: Fluorescence-based RBDCC screen. One pool of resin (pool 7) from the screen exhibits significant fluorescence, and as such a small-molecule ligand for the target DNA. Right: Selected homodisulfide resin-bound dynamic combinatorial library member that selectively binds DNA Sequence 2.

3.3. RNA as a Target for DCC

Ribonucleic acid is no longer viewed simply as an intermediary between the genetic code of DNA and proteins; we live in an "RNA world" [24]. RNAs are involved in a plethora of cellular processes beyond their familiar role as messenger, including transcriptional and translational regulation, and enzymatic catalysis. Even as a "messenger", the complex tertiary structure of RNA makes it an intriguing target for small-molecule intervention, and an important challenge to chemists' ability to design selective binders. However, RNA has proven more difficult to selectively target with small molecules than its DNA counterpart. Selective and high affinity small-molecule recognition of RNA is currently a large and growing area of bioorganic chemical research, as small-molecule RNA binding can have great effects on translation and cellular function [25]. This section will discuss dynamic approaches to the development of small-molecule ligands for RNA sequences.

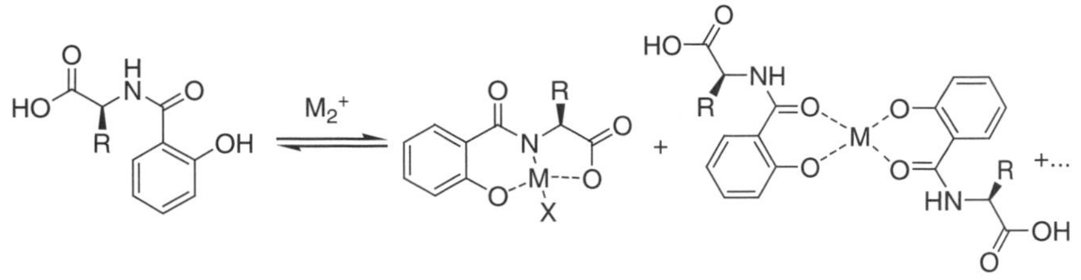

Figure 3.12   Potential coordination modes for metal salicylamides.

Building on their earlier salicylamide–metal-based dynamic combinatorial methods, Karan and Miller targeted an RNA hairpin sequence modeled after the GRP-binding P7 helix of the Group I intron of *Pneumocystis carinii* [26]. Library selection was performed by equilibrium dialysis, where the dialysis tubing-encapsulated RNA hairpin sequence (5′-UAGUCUUUC-GAGACUA-3′) was immersed in a salicylamide and Cu^{2+} solution. In this approach, the concentration of small-molecule ligands binding the RNA is enhanced in the dialysis tubing, while on-binding compounds remain at equal concentration on both sides of the tubing. After equilibration, the small-molecule RNA ligands are eluted from the dialysis tubing by a series of dialyses against buffer or water, dried, and analyzed by HPLC. In this fashion, a library of salicylamide Cu^{2+} complexes was screened against the RNA hairpin target and an analogous DNA hairpin sequence (Figs. 3.12 and 3.13). In some respects, this is a "fully solution phase" analog of the affinity reagent method used earlier by the Miller group in the identification of DNA-binding compounds, and by Eliseev and coworkers in the identification of guanidinium binders. Like those examples, the equilibrium dialysis method provides a way to physically separate the bulk of the equilibrating small-molecule library from the target. Furthermore, the multiple dialysis steps were intended to provide further amplification through library replenishment (although in practice the success of such multicycle methods depends on the dissociation constant of the selected compound(s); this has been discussed further in a theoretical context by Severin [27]).

The equilibrium dialysis experiment revealed that histidine-substituted salicylamide was selected as an RNA ligand. Subsequent binding analysis by UV titrations and Job plot revealed the histidine-substituted salicylamide Cu^{2+} complex bound the target RNA hairpin with an apparent dissociation constant of 150 nM. This binding constant likely reflects more complex binding processes than a simple 1:1 interaction, as the observed binding curve saturates well below the concentration of the histidine-substituted salicylamide, and thus the actual affinity of the complex for targeted RNA is probably lower. Importantly, however, titrations with the

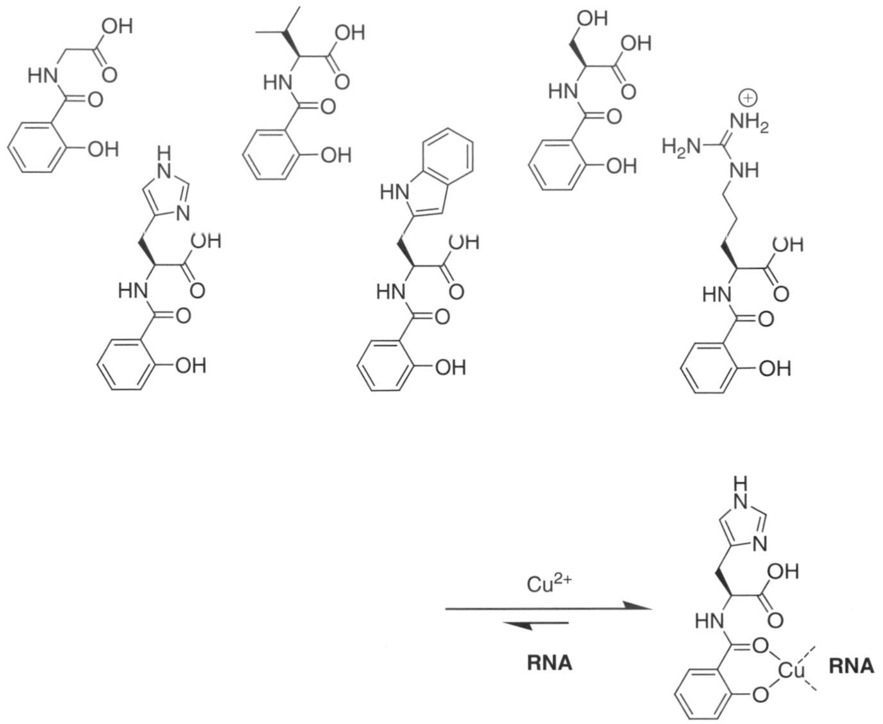

Figure 3.13   Dynamic salicylamide–copper complexation identifies a histidine-substituted salicylamide that selectively binds an RNA hairpin based on the Group I intron of *P. carinii* over the analogous DNA hairpin.

analogous DNA hairpin showed the histidine-substituted salicylamide Cu^{2+} complex exhibited greater than 300-fold selectivity for the RNA hairpin. This example proved the utility of DCC for the development of novel RNA-binding compounds. As in the salicylaldimine–zinc libraries targeting DNA described previously, it also highlighted some of the challenges associated with the use of labile metal complexation as an exchange reaction for DCC experiments.

Building on these initial observations and lessons, Miller and coworkers expanded the RBDCC approach used previously in the identification of DNA-binding compounds, by developing an 11,325-member library targeting RNA (Fig. 3.14). This library was first screened against an HIV1 frameshift-inducing RNA stemloop [28]. Given that approximately 39.5 million people are infected with the human immunodeficiency virus (HIV) worldwide [29], HIV-targeted therapies continue to be an area of intense pharmaceutical research. The frameshift-inducing RNA stemloop targeted in this DCL is involved in the regulation of the expression of the HIV proteins Gag and Gag-Pol, and as such is therefore a potentially significant

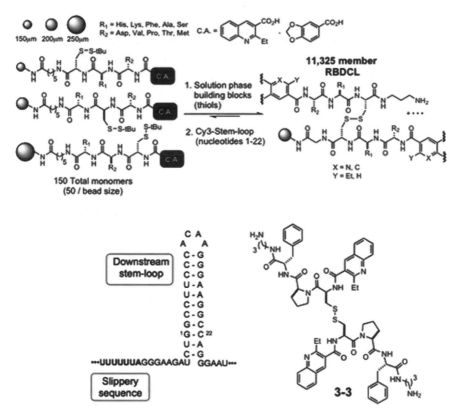

Figure 3.14 Top: Design of the 11,325-member resin-bound dynamic combinatorial library (RBDCL). Bottom Left: The HIV1 frameshift-inducing stemloop RNA screened against the large RBDCL. Bottom Right: Compound **3–3**, the homodisulfide selected from the RBDCL that binds the HIV1 frameshift-inducing stemloop with high affinity and selectivity.

therapeutic target [30–32]. Expanding on the initial design of the first RBDCC report, a library of N-terminal aromatic capped, cysteine containing tripeptides capable of disulfide exchange was synthesized. The resin-bound approach was chosen to overcome common LC-MS hurdles associated with deconvolution of large DCLs. As mentioned earlier, the RBDCL experimental protocol consists of screening the RBDCL against a labeled target, in this case a Cy-3-labeled HIV1 frameshift-inducing RNA stemloop. Any selected RBDCL member that binds the fluorescent target is easily fluorescently visualized, and can be spatially segregated allowing facile identification of the selected DCL member. Chemical diversity in the 150 unique thiol-containing building blocks was achieved by incorporating a variety of different amino acids and carboxylic acids, in addition to varying the position of the cysteine residue in the sequence. The 150 unique thiols produce an 11,325-member disulfide DCL. The library was screened against the fluorescently labeled HIV1 frameshift-inducing RNA stemloop and three resin-bound library members exhibited significant fluorescence, indicating RNA target binding. The selected library members were cleavage from the resin and identified by mass spectrometry (MS). Experimental conditions allowed for only the monothiol disulfide precursors from the three selected resin beads to be identified. Therefore, it was determined that only disulfides from these three monothiol precursors were selected ligands for the HIV1 frameshift-inducing RNA stemloop. Secondary screening of the nine possible disulfide ligands suggested that compound **3–3** was the best binder of those selected.

Subsequent binding analysis revealed the selected RBDCL member **3–3** binds the target HIV1 frameshift-inducing stemloop with good affinity (K_D = 4.1 µM by SPR; K_D = 350 nM by fluorescence titration) and significant selectivity over alternate nucleic acid sequences. This novel RBDCC approach allows efficient screening of large DCLs, producing a high affinity and selective ligand for an important RNA sequence involved in the HIV1 life cycle. While the disulfide is likely to have limited biostability, it will be interesting to see if compounds related to **3–3** but with less labile groups replacing the disulfide will have any biological effects on the HIV1 frameshift process.

The potential of this large RBDCL to address other RNA sequences was recently demonstrated by the Miller group, who screened it against an RNA believed to be the primary cause of Type 1 myotonic dystrophy (DM1) [33]. This disease, the most common form of adult-onset muscular dystrophy, is thought to result from the uncontrolled expansion of CTG repeats in the *DMPK* (DM protein kinase) gene. The $(CUG)_n$ RNA transcribed from these repeats has been found to sequester splicing factors in the nucleus, including a protein dubbed Muscleblind. Screening the 11,325-member RBDCL against 5'-CCG-$(CUG)_{10}$-CGG-3' RNA allowed identification of compounds **4–7** (Fig. 3.15) as components of putative binders. Further analysis verified that several of the 10 possible unique disulfides formed from these monomers bound target (CUG) repeat RNA with low-micromolar dissociation constants (K_D), and, most importantly, also inhibited binding of MBNL-1 to (CUG) repeat RNA with IC_{50} values similar to their K_D values.

3.4. Beyond Binding: Nucleic Acids as Components of DCC Experiments

The previous sections have detailed experiments focused on the use of DCC to identify new organic compounds capable of binding "biologically

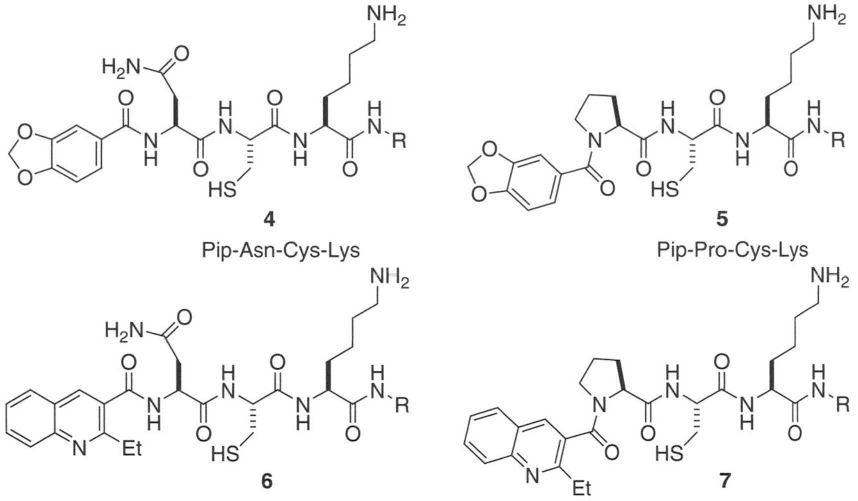

4
Pip-Asn-Cys-Lys

5
Pip-Pro-Cys-Lys

6
Quin-Asn-Cys-Lys

7
Quin-Pro-Cys-Lys

R = (CH$_2$)$_3$NH$_2$ or (CH$_2$)$_5$C(O)NH$_2$

Figure 3.15   Components of (CUG)$_n$ RNA-binding compounds identified by RBDCC.

interesting" nucleic acids. However, several groups have demonstrated that the concepts behind DCC can impact nucleic acids in other ways. One can also use dynamic combinatorial methodology to *modify* nucleic acids, as well as bind to them. Dynamic modification of nucleic acids has been employed to stabilize DNA and RNA structures, and to modify bases and nucleic acids to yield novel supramolecular polymers and hydrogels.

In 2004, Rayner and coworkers reported a dynamic system for stabilizing nucleic acid duplexes by covalently appending small molecules [34]. These experiments started with a system in which 2-amino-2′-deoxyuridine (U-NH$_2$) was site-specifically incorporated into nucleic acid strands via chemical synthesis. In the first example, U-NH$_2$ was incorporated at the 3′ end of the self-complementary U($-$NH$_2$)GCGCA DNA. This reactive amine-functionalized uridine was then allowed to undergo imine formation with a series of aldehydes (**Ra–Rc**), and aldehyde appendages that stabilize the DNA preferentially formed in the dynamic system. Upon equilibration and analysis, it was found that the double-stranded DNA modified with nalidixic aldehyde **Rc** at both U-NH$_2$ positions was amplified 34% at the expense of **Ra** and **Rb** (Fig. 3.16). The **Rc**-appended DNA stabilizing modification corresponded to a 33% increase in T_m (melting temperature). Furthermore, imine reduction of the stabilized DNA complex with NaCNBH$_3$ resulted in a 57% increase in T_m.

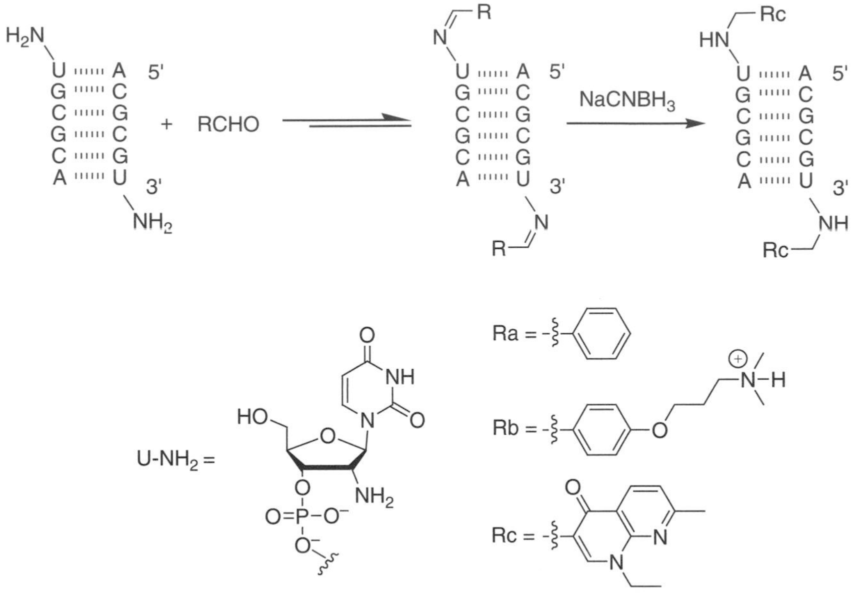

Figure 3.16 Dynamic combinatorial modification of 2′-amino-2-deoxyuridine (U-NH$_2$) containing DNA by aldehydes **Ra–Rc**. The nalidixic aldehyde was selectively appended to both U-NH$_2$ positions resulting in DNA duplex stabilization.

Similarly, the authors also examined the stabilization effect of dynamic modification of a U-NH$_2$-appended RNA aptamer that forms a kissing complex with the HIV1 transactivation-responsive RNA element TAR. In this dynamic library, 2-chloro-6-methoxy-3-quinolinecarboxaldehyde (**Rd**) was incorporated in place of benzaldehyde (**Ra**). After equilibration of the U-NH$_2$-substituted aptamer and aldehydes **Rb–Rd** in the presence of the TAR RNA target, it was found that the nalidixic aldehyde **Rc**-appended RNA was amplified 20%, and accompanied by an increased T_m (Fig. 3.17). Interestingly, the nalidixic aldehyde **Rc** was selected in both DNA and RNA complexation experiments.

Building on these initial results, the Rayner group next investigated stabilization effects upon modification of two U-NH$_2$ groups in a 10-nucleotide (nt) RNA duplex [35]. This library allowed for 15 possible modifications (6 mono- and 9 di-functionalized). As in their earlier work, it was found that the nalidixic aldehyde preferentially formed imino-conjugated oligonucleotides (Fig. 3.18). Deconvolution by MS and HPLC showed that the RNA complement containing no modification at U-NH$_2$ at position 3 and nalidixic aldehyde conjugate at U-NH$_2$ at position 9 was amplified to the greatest extent and accompanied by a modest increase in T_m.

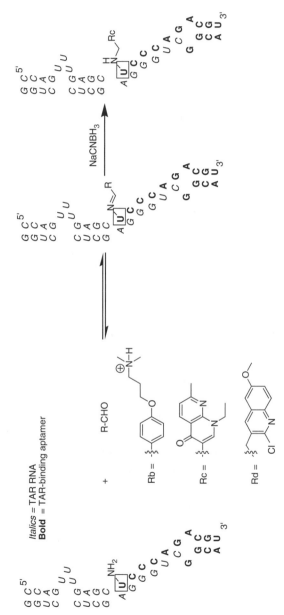

Figure 3.17 Approach to the dynamic combinatorial modification of the TAR-binding aptamer. Left, italics: The TAR RNA sequence. Left, bold: The TAR-binding aptamer. Left, boxed: The 2′-amino-2-deoxyuridine (U-NH₂) for dynamic RNA modification. Left center: **Rb—Rd**, the aldehyde library components. Right center: Imino-linked DCL members. Right: The selected nalidixic aldehyde appended to U-NH₂ results in the TAR RNA–aptamer complex stabilization.

103

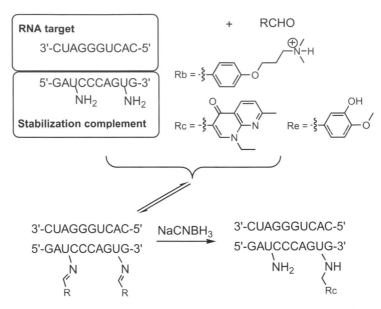

Figure 3.18   Approach to the dynamic combinatorial modification of RNA.

The results of these three dynamic systems highlight the ability of DCC to identify the sites and appendages that most efficiently stabilize nucleic acid interactions.

Rayner and coworkers recently demonstrated the incorporation of this DCC nucleic acid appending strategy into a systematic evolution of ligands by exponential enrichment (SELEX) *in vitro* selection approach [36]. SELEX has found great utility in the development of nucleic-acid-based aptamers, which bind with high affinity and selectivity to targets of interest [37]. However, SELEX has traditionally been limited to using standard nucleic acids. The ability to chemically modify the nucleic acids during the SELEX process can greatly increase the chemical diversity of the system. To this end, the appended nucleic acid DCC approach was incorporated into a SELEX system to develop modified RNA aptamers targeting the TAR element of HIV1. This DCC SELEX system is outlined in Fig. 3.19.

In the DCC SELEX screen, aldehydes, the TAR RNA target, and a random library of 2′-amino RNAs were allowed to equilibrate. Next, the TAR RNA target and bound ligands were separated from the aptamer library. The selected 2′-amino RNAs that bound the TAR RNA target were then reverse transcribed into DNA and PCR amplified. These double-stranded

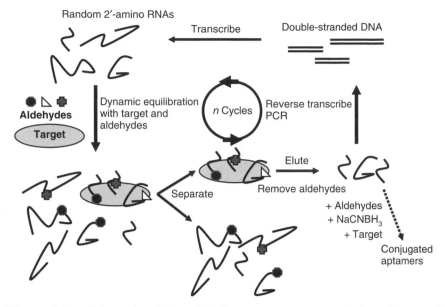

Figure 3.19 Schematic of the DCC SELEX system. Upper left: A library of random 2'-amino RNAs are allowed to equilibrate via imine formation with aldehydes in the presence of target. Bottom left: Modified RNAs are bound to the target. Bottom center: Modified RNAs bound to the target are separated from unbound RNAs. Bottom right: Selected RNAs are eluted and reverse transcribed and amplified to corresponding double-stranded DNA. Upper right: The selected double-stranded DNA is transcribed to the 2'-amino RNAs. The selection process is repeated n-cycles and selected conjugated aptamers are identified.

DNAs were then transcribed into the corresponding 2'-amino RNAs and the selection process was repeated.

The TAR RNA target sequence, the 2'-amino RNA library and the appended aldehydes were subjected to the DCC SELEX system. The screen selected a 19-nt sequence with U-NH$_2$ appended at position 9 and unmodified at positions 6 and 7 (Fig. 3.20). Importantly, it was shown that different sequences were identified when control selections were carried out in the absence of aldehydes, proving that the imino-conjugated nucleic acids are being selected.

Subsequent analysis showed the modified aptamers bound the target TAR RNA with high (10^{-8} M) affinity and increased the stability of the aptamer–RNA complex. However, it does not seem that the identity of the appendage at position U-NH$_2$ 9 greatly affects binding affinity. Nevertheless, this proof-of-principle multidimensional system demonstrated

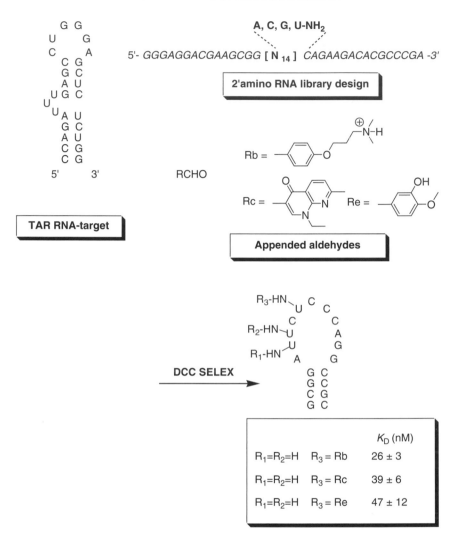

Figure 3.20   TAR RNA DCC SELEX system, employing 2′-amino-2-deoxyuridine (U-NH$_2$) capable of reversible imine formation with the appended aldehydes **Rb**, **Rc**, and **Re**. Selected appended RNA aptamers and their corresponding dissociation constants are shown at the bottom.

that DCC applied with SELEX holds promise to expand the diversity and affinity of SELEX-selected aptamers, and it will be interesting to see how the application of this approach furthers the field.

In addition to stabilizing appendages, dynamic nucleic acid modification has been applied in supramolecular chemistry to develop novel materials. To this end, Lehn and coworkers developed dynamic nucleic acid analogs,

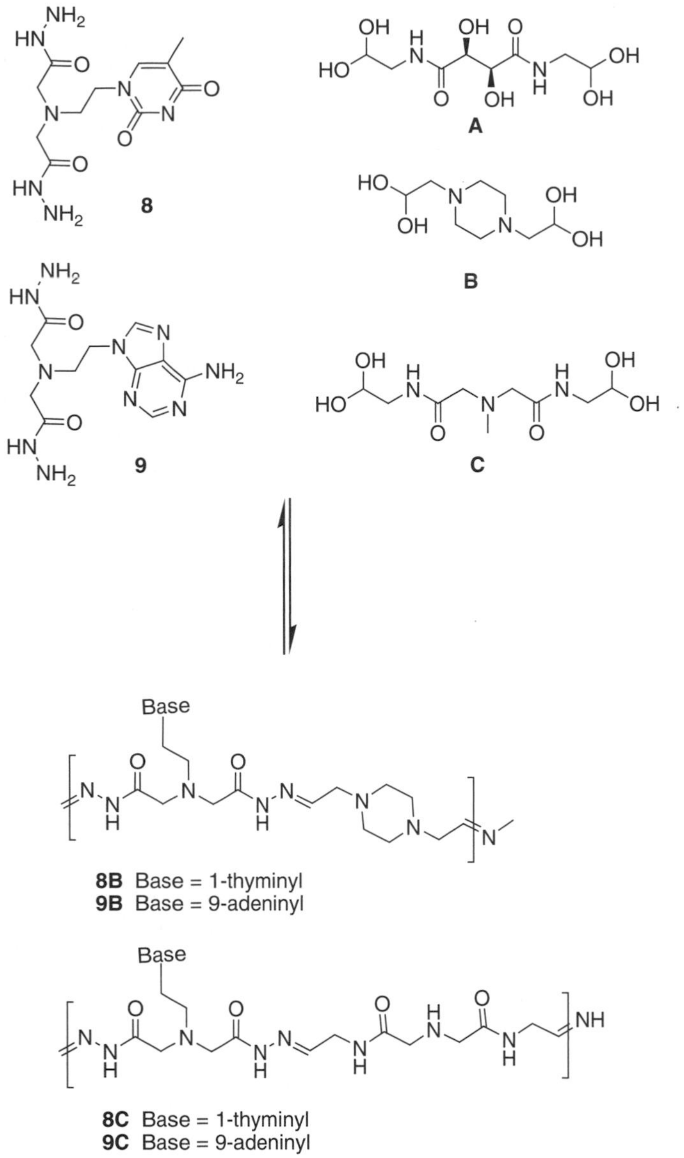

Figure 3.21   The design of dynamic nucleic acid analogs **8** and **9**, or DyNAs, for dynamic acylhydrazone formation with **A**–**C** to develop polycationic polymers.

or DyNAs [38]. These nucleobase derivatives were found to be capable of forming dynamic polycationic polymers. In this system, acylhydrazone formation between nucleobase dihydrazides **8** and **9** and hydrated dialdehydes **A**–**C** under aqueous conditions produced various polycationic polymers (Fig. 3.21). Condensation and polymerization were monitored by

^{1}H NMR and multiangle laser light scattering. Library screening showed varying degrees of polymerization, with the highest degree of polymerization corresponding to **8C**. All polymerization trials with compound **A** resulted in precipitation. Therefore, further work was focused on polymerizations with compounds **B** and **C**.

The novel positively charged polymers were hypothesized to have molecular recognition capabilities for negatively charged species. Indeed, polycationic polymers **8B** and **9B** were shown to electrostatically interact with poly-dA DNA. Furthermore, the adaptive and dynamic features of the polymerization of **8** and **B** in the presence of anionic targets were investigated. Dynamic polymerization was performed in the presence of anionic targets such as inositol hexaphosphate (IHP), inositol tripyrophosphate (ITPP), polyaspartic acid, and ATP. These experiments revealed that the degree of polymerization of **8B** (a) increased with degree of negative charge on the target, ATP(-4) < ITPP(-6) < IHP(-12); (b) increased with molecular weight of anionic target; and (c) was saturable at high target concentration.

Recently Lehn and coworkers investigated the supramolecular properties of G-quartet assemblies upon dynamic covalent decoration [39]. Previously the Lehn group studied guanosine-5′-hydrazide hydgrogel formation in the presence of metal ions [19]. The hydrazide moity provides an excellent dynamic diversification point for decoration with aldehyde substituents. In this study, pyridoxal monophosphate reversibly formed acylhydrazone bonds with the guanosine-5′-hydrazide quartets in the presence of Na$^+$ or K$^+$, and the hydrogel characteristics were analyzed with small-angle neutron scattering (Fig. 3.22).

G-hydrazide alone self-assembles to form fibers in the presence of Na$^+$ or K$^+$. The fibers are much thicker in the presence of Na$^+$ than in K$^+$, presumably due to differences in the size of the cation. Interestingly, when the G-hydrazide was dynamically appended with pyridoxal monophosphate, it formed loosely stacked single-column quartets, presumably due to the negatively charged appended phosphate groups inhibiting lateral aggregation and producing a more viscoelastic solution. The results of this dynamic assembly lend insight into the properties of G-hydrazide hydrogels, and may be of future use in the design of improved hydrogels.

3.5. DCC–Nucleic Acid Templating

Dynamic combinatorial chemistry has recently been used to investigate the process of templating nucleic acid library members to form interesting structures. To date, the templating process has included the development

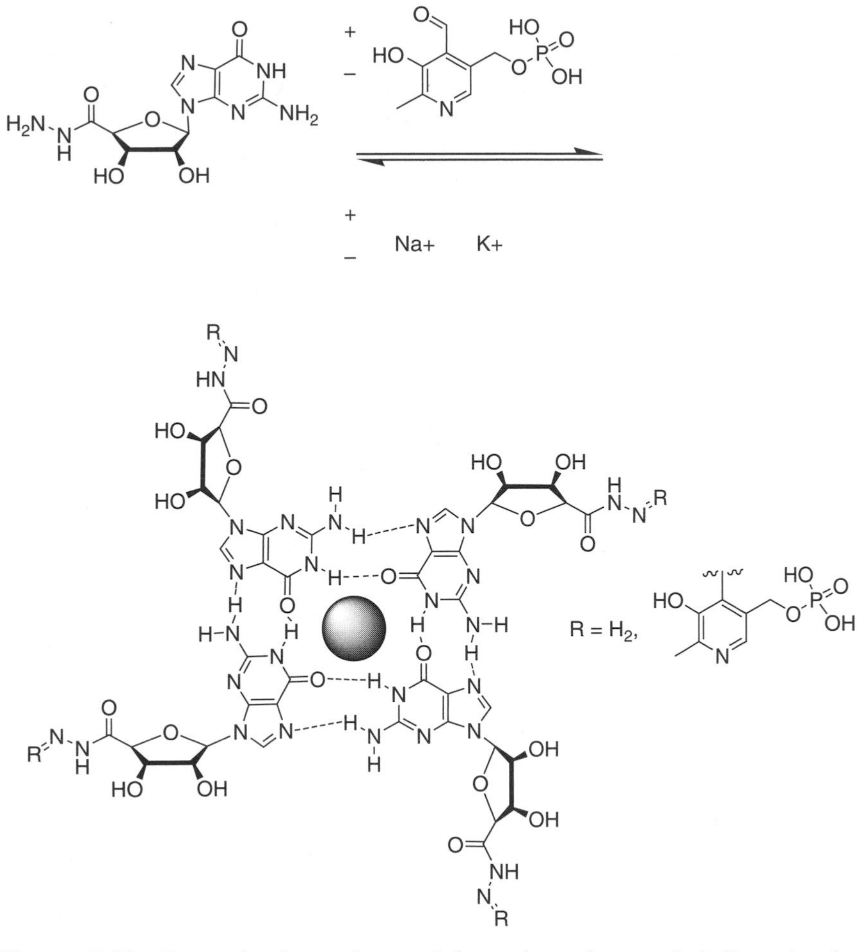

Figure 3.22 Dynamic decoration and formation of appended G-quadruplex structures capable of forming hydrogels.

of DNA nanostructures, dynamic assembly of DNA primers, selection of enantioselective nucleoside receptors, and formation of PNA quadruplexes and metallocalixarenes. As discussed in Chapter 1, spontaneous assembly of nucleotides on nucleic acid templates, a key experimental focus of early origins-of-life research in the 1980s and early 1990s, was an important precursor to the development of DCC. Dynamic assembly of DNA on target single-stranded DNA sequences was described by Zhan and Lynn in 1997 [40], building on work reported by Goodwin and Lynn in 1992 [41]. In this system, imine formation between 5′ amine functionalized and 3′ aldehyde functionalized trinucleotides allowed for dynamic assembly of duplex DNA. The templated strand contains an unnatural imine linkage in place of the natural phosphate linkage. In this proof-of-principle experiment, 5'-dCGT-CH$_2$CHO and H$_2$N-dTGC-3' were allowed to equilibrate in the presence of the complementary dGCAACG sequence (Fig. 3.23). Reaction kinetics were monitored by NMR, and proved the equilibration was under thermodynamic control. The templated imine–oligo duplex was shown to have a stability similar to the native duplex as evidenced by NMR melting curves, thus proving the utility of DCC in complementary DNA templating reactions.

One can easily envision the utility of templated DNA assembly applied to selective primer formation for polymerase reactions and genetic expression. However, this approach is potentially problematic, as the unnatural internal imine linkage in the templated oligo may interfere with native polymerase activity. In 2006 Benner and coworkers reported a system in which unnatural DNA primers to complementary DNA target strands were formed by dynamic imine formation, and were shown to be suitable substrates for polymerases [42]. The rationale behind the dynamic primer assembly was to increase primer specificity. It has been shown that on average a 16-nt primer will have a single complementary sequence in the human genome. However, in some instances, mismatches in primer-mediated duplex formation occur, permitting unwanted priming and polymerase products. On the other hand, shorter oligos, consisting of 6–8 nt, do not tolerate such mismatches and are able to form much more stable duplexes. Unfortunately, these short oligos lack specificity and have many complementary DNA sequences in the human genome (~1 × 10^6 for a 6-nt oligo). Thus, a dynamic system to develop composite primers could exploit the discrimination of short oligos and specificity of longer oligos. In this system, a 5'-d(8 nt)-CHO was incubated with H$_2$N-d(6 or 8 nt) in the presence of target DNA. The selected primers were then tested for polymerase primer extension activity. Indeed, it was found that 9°N DNA polymerase efficiently elongated the imine-linked 8 nt + 6 nt composite primer selectively and efficiently. This primer formation approach represents an interesting example of dynamic assembly applied to templating nucleic acids.

Figure 3.23 Dynamic combinatorial assembly of DNA primers on a template through imine formation between 5'-amino-substituted and 3'-aldehyde-substituted DNA strands.

In recent years, DNA has emerged as a valuable building block in emerging applications in nanoscience [43]. When combined with DCC, one can begin to create very interesting nanoscale structures. For example, Aldaye and Sleiman employed a dynamic combinatorial approach to selectively drive the formation of specific DNA nanostructures (Fig. 3.24) [44]. The library consisted of two organic scaffolds identically substituted at two or three positions with single-stranded DNA; DNA on each scaffold was complementary to the other scaffold. These scaffolds were allowed to reversibly hybridize to form double-stranded DNA-linked assemblies. In the absence of templating guest, a complex set of DNA networks was observed. For the disubstituted scaffolds, formation of the library in the presence of Ru(bpy)$_3^{2+}$ initially produced a complex series of cyclic and linear oligomers, which then "evolved" to the dimer. Analogous templated evolution of the library based on trisubstituted scaffolds led to the preferential formation

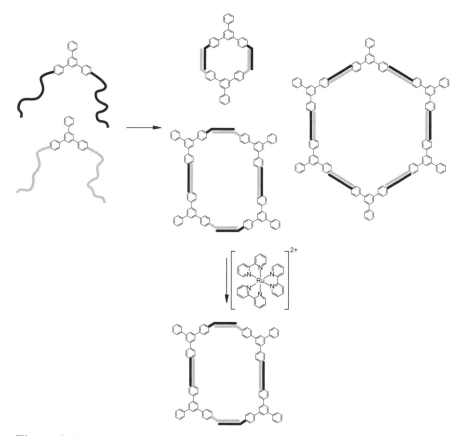

Figure 3.24 Dynamic combinatorial selection of DNA nanostructure (Aldaye and Sleiman).

of single ladder-like nanostructures. At longer time points, these ladders formed large supramolecular DNA fibers greater than 50 μm in length as evidenced by atomic force microscopy. The vast numbers of DNA scaffolds and DNA-binding molecules that could be applied to similar dynamic assembly systems hold great promise for the technique's utility in the tunable development of novel DNA nanostructures.

In another application of dynamic nucleic acid templating, Balasubramanian and coworkers studied the effects of quadruplex formation in dynamic protein nucleic acid (PNA) oligomers [45]. Two PNA thiol-containing building blocks, **T** and **G** (**T** = Lys-TTTT-Gly-Gly-Cys, **G** = Lys-TGGG-Gly-Gly-Cys), were allowed to equilibrate via disulfide exchange to produce the three possible disulfide library members **T–T**, **T–G**, and **G–G**. It was hypothesized that prearrangement of **G** would form a templating quadruplex motif, and affect the dynamic disulfide exchange process. Indeed, the **G**-quadruplex formation enhanced the formation of the **G–G** complex. Through a series of deuterium exchange, oxidation, ion dependence, and temperature screens, the quadruplex motif was determined to form prior to disulfide formation. This prearrangement acts as a "thermodynamic sink" templating element and drives the dynamic enhancement of the **G–G** product.

Recently, enantioselective receptors templated on (–)-adenosine have been developed in a dynamic combinatorial fashion by Gagné and coworkers [46]. In their system, they screened a racemic proline dipeptide hydrazone exchange DCL to identify receptors for enantiopure guest (–)-adenosine (Fig. 3.25). Enantiomeric excess was monitored by HPLC-laser polarimetry and MS. It was found that the dimer species was amplified greater than higher order oligomers, and that the SS dimer enantiomer was selected. The elegant combination of laser polarimetry and pseudo-enantiomer methodology allowed the direct stereochemical identification of the selected library member from a complex racemic library. Further discussion of this approach is provided in Chapter 5.

Additionally, nucleic acid bases have been used in the dynamic assembly of mixed-metal, mixed-pyrimidine metallacalix[n]arenes [47]. In this approach, Lippert and coworkers investigated the dynamic assembly of metallacalixarenes based on platinum (Pt^{II}), palladium (Pd^{II}), uracil, and cytosine assemblies with mixed amines. These combinations form cyclic metallacalix[n]arenes structures with $n = 4$ and $n = 8$. Of the metallacalix[4]arenes, compounds were formed with five, six, and eight bonded metals, and a variety of nucleobase connectivities (UCUC and UCCU). The dynamic nature of this assembly allows access to novel and structurally diverse set of nucleobase metallacalixarenes.

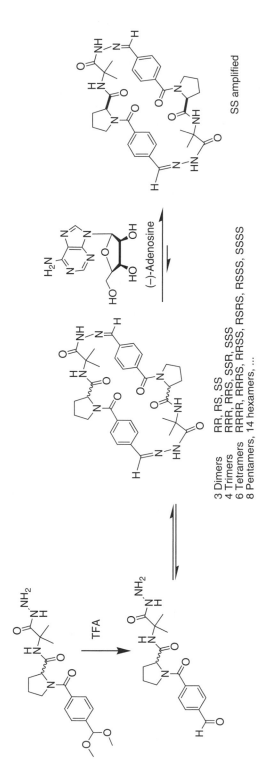

3 Dimers RR, RS, SS
4 Trimers RRR, RRS, SSR, SSS
6 Tetramers RRRR, RRRS, RRSS, RSRS, RSSS, SSSS
8 Pentamers, 14 hexamers, ...

Figure 3.25 Racemic dynamic combinatorial library targeting (−)-adenosine. Left: Racemic porline containing dipeptide library building block. Middle: Various enantiomeric oligomers formed through reversible hydrazone exchange. Right: The SS enantiomer selected upon equilibration with (−)-adenosine.

3.6. Conclusions

The examples presented in this chapter show that DCC can indeed serve as a useful method for rapidly identifying compounds capable of binding DNA and RNA targets of interest. Furthermore, DCC concepts can be applied to nucleic acid modification, allowing the "evolution" of oligonucleotides with desirable physical properties. While the question arises as to whether DCC screening methods provide the "best" answer in all cases (e.g., Is the tightest binder in the library identified unambiguously? A question examined further in the theoretical studies discussed in Chapter 1), these methods unquestionably provide a rapid method for lead compound discovery. The resin-bound form of DCC (RBDCC) has shown that screening of relatively large libraries is possible, and work in the authors' laboratory is ongoing to extend this methodology to a broad range of targets and compound types. Finally, early applications of DCC in DNA nanostructure synthesis to allow ready access to well-defined nanomaterials suggest that this will become a vibrant area of research in itself.

The short history of nucleic-acid-targeted DCC to date also provides several lessons in library and experimental design that must be carefully considered when embarking on a DCL project. As with all DCC screens, the scrambling reaction must be compatible with the biological target in question. In some respects, this is a simpler challenge for nucleic acid targeting than it is for other biopolymer targets; DNA and RNA generally do not have reactive nucleophilic amines. However, reversible reactions must obviously operate in an aqueous environment, and must be compatible with the relatively high salt conditions required for DNA and RNA secondary structural stability. These reactions must also allow for stopping the equilibration, as a prerequisite to analyzing the outcome of the screen. Thus far, it appears that disulfide exchange is the method of choice of the reactions examined, although imine and acylhydrazone reactions have also proven successful. Metal complexation has been useful in "proof-of-concept" experiments, but is complicated both by the promiscuous coordination preferences of some metal ions and by difficulties associated with halting the equilibration. We can expect that future efforts will provide additional reactions to add to these.

References

1. Dervan, P. B.; Bürli, R. W. Sequence-specific DNA recognition by polyamides. *Curr. Opin. Chem. Bio.* **1999**, *9*, 77–91.
2. Tse, W.; Boger, D. Sequence-selective DNA recognition: Natural products and nature's lessons. *Chem. Biol.* **2004**, *11*, 1607–1617.

3. Dawson, S.; Makinson, J. P.; Paumier, D.; Searcey, M. Bisintercalator natural products with potential therapeutic applications: Isolation structure determination, synthetic and biological studies. *Nat. Prod. Rep.* **2007**, *24*, 109–126.

4. Dervan, P. B.; Doss, R. M.; Marques, M. A. Programmable DNA binding oligomers for control of transcription. *Curr. Med. Chem. Anticancer Agents* **2005**, *5*, 373–387.

5. Dervan, P. B. Molecular recognition of DNA by small molecules. *Bioorg. Med. Chem.* **2001**, *9*, 2215–2235.

6. Klekota, B.; Hammond, M. H.; Miller, B. L. Generation of novel DNA-binding compounds by selection and amplification from self-assembled combinatorial libraries. *Tetrahedron Lett.* **1997**, *38*, 8639–8642.

7. Garnovskii, A. D.; Nivorozhkin, A. L.; Minkin, V. I. Ligand environment and the structure of schiff base adducts and tetracoordinated metal-chelates. *Coord. Chem. Rev.* **1993**, *126*, 1–69.

8. Klekota, B.; Miller, B. L. Selection of DNA-binding compounds *via* multi-stage molecular evolution. *Tetrathedron* **1999**, *55*, 11687–11697.

9. Eliseev, A. V.; Nelen, M. I. Use of molecular recognition to drive chemical evolution. 1. Controlling the composition of an equilibrating mixture of simple arginine receptors. *J. Am. Chem. Soc.* **1997**, *119*, 1147–1148.

10. Whitney, A. M.; Ladame, S.; Balasubramanian, S. Templated ligand assembly by using G-quadruplex DNA and dynamic covalent chemistry. *Angew. Chem. Int. Ed.* **2004**, *43*, 1143–1146.

11. Oganesian, L.; Bryan, T. M. Physiological relevance of telomeric G-quadruplex formation: A potential drug target. *Bioessays* **2007**, *29*, 155–165.

12. Cuesta, J.; Read, M. A.; Neidle, S. The design of G-quadruplex ligands as telomerase inhibitors. *Mini Rev. Med. Chem.* **2003**, *3*, 11–21.

13. Ladame, S.; Whitney, A.; Balasubramanian, S. Targeting nucleic acid secondary structures with polyamides using an optimized dynamic combinatorial approach. *Angew. Chem. Int. Ed.* **2005**, *44*, 5736–5739.

14. Nielsen, M. C.; Ulven, T. Selective extraction of G-quadruplex ligands from a rationally designed scaffold-based dynamic combinatorial library. *Chem. Eur. J.* **2008**, *14*, 9487–9490.

15. McNaughton, B. R.; Miller, B. L. Resin-bound dynamic combinatorial chemistry. *Org. Lett.* **2006**, *8*, 1803–1806.

16. Hioki, H.; Still, W. C. Chemical evolution: A model system that selects and amplifies a receptor for the tripeptide (D)Pro(L)Val(D)Val. *J. Org. Chem.* **1998**, *63*, 904–905.

17. Miller, B. L.; Klekota, B. U.S. Patent #6,599,754, **2003**.

18. Pérez-Fernández, R.; Pittelkow, M.; Belenguer, A. M.; Sanders, J. K. M. Phase-transfer dynamic combinatorial chemistry. *Chem. Commun.* **2008**, 1738–1740.

19. Sreenivasachary, N.; Lehn, J.-M. Gelation-driven component selection in the generation of constitutional dynamic hydrogels based on guanine-quartet formation. *Proc. Natl. Acad. Sci. U.S.A.* **2005**, *102*, 5938–5943.

20. McNaughton, B. R.; Miller, B. L. A mild and efficient one-step synthesis of quinolines. *Org. Lett.* **2003**, *523*, 4257–4259.

21. Dervan, P. B.; Van Dyke, M. M. Echinomycin binding sites on DNA. *Science* **1984**, *225*, 1122–1127.

22. Addess, K. J.; Sinsheimer, J. S.; Feigon, J. Solution structure of a complex between [N-MeCys^3,N-MeCys^7]TANDEM and [d(GATATC)]$_2$. *Biochemistry* **1993**, *32*, 2498–2508.

23. Ladame, S. Dynamic combinatorial chemistry: On the road to fulfilling the promise. *Org. Biomol. Chem.* **2008**, *6*, 219–226.

24. Gilbert, W. Origin of life: The RNA world. *Nature* **1986**, *319*, 618.

25. Thomas, J. R.; Hergenrother, P. J. Targeting RNA with small molecules. *Chem. Rev.* **2008**, *108*, 1171–1224.

26. Karan, C.; Miller, B. L. RNA-selective coordination complexes identified via dynamic combinatorial chemistry. *J. Am. Chem. Soc.* **2001**, *123*, 7455–7456.

27. Severin, K. The advantage of being virtual – Target-induced adaptation and selection in dynamic combinatorial libraries. *Chem. Eur. J.* **2004**, *10*, 2565–2580.

28. McNaugton, B. R.; Gareiss, P. C.; Miller, B. L. Identification of a selective small-molecule ligand for HIV-1 frameshift-inducing stem-loop RNA from an 11,325 member resin bound dynamic combinatorial library. *J. Am. Chem. Soc.* **2007**, 129(*37*), 11306–11307.

29. http://www.unaids.org/en/HIV_data/2006GlobalReport/default.asp.

30. Biswas, P.; Jiang, X.; Pacchia, A. L.; Dougherty, J. P.; Peltz, S. W. The human immunodeficiency virus type 1 ribosomal frameshifting site is an invariant sequence determinant and an important target for antiviral therapy. *J. Virol.* **2004**, *78*, 2082–2087.

31. Dulude, D.; Brakier-Gingras, L. The structure of the frameshift stimulatory signal in HIV-1 RNA: A potential target for the treatment of patients with HIV. *Med. Sci. (Paris)* **2006**, *11*, 969–972.

32. Hung, M.; Patel, P.; Davis, S.; Green, S. R. Importance of ribosomal frameshifting for human immunodeficiency virus type 1 particle assembly and replication. *J. Virol.* **1998**, *72*, 4819–4824.

33. Gareiss, P. C.; Sobczak, K.; McNaughton, B. R.; Thornton, C. A.; Miller, B. L. Dynamic Combinatorial Selection of Small Molecules Capable of Inhibiting the (CUG) Repeat RNA – MBNL1 Interaction in vitro: Discovery of Lead Compounds Targeting Myotonic Dystrophy (DM1). *J. Am. Chem. Soc.* **2008**, *130*, 16524–16261.

34. Bugaut, A.; Toulmé, J.-J.; Rayner, B. Use of dynamic combinatorial chemistry for the identification of covalently appended residues that stabilize oligonucleotide complexes. *Angew. Chem. Int. Ed.* **2004**, *43*, 3144–3147.

35. Bugaut, A.; Bathany, K.; Scmitter, J.-M.; Rayner, B. Target-induced selection of ligands from a dynamic combinatorial library of mono- and bi-conjugated oligonucleotides. *Tetrahedron Lett.* **2005**, *46*, 687–690.

36. Bugaut, A.; Toulmé, J-J.; Rayner, B. SELEX and dynamic combinatorial chemistry interplay for the selection of conjugated RNA aptamers. *Org. Biomol. Chem.* **2006**, *4*, 4082–4088.

37. Stoltenburg, R.; Reinemann, C.; Strehlitz, B. SELEX – A (r)evolutionary method to generate high-affinity nucleic acid ligands. *Biomol. Eng.* **2007**, *24*, 381–403.

38. Sreenivasachare, N.; Hickman, D. R.; Sarazin, D.; Lehn, J.-M. DyNAs: Constitutional dynamic nucleic acid analogues. *Chem. Eur. J.* **2006**, *12*, 8581–8588.

39. Buhler, E.; Sreenivasachary, N.; Candau, S.-J.; Lehn, J.-M. Modulation of the supramolecular structure of G-quartet assemblies on dynamic covalent decoration. *J. Am. Chem. Soc.* **2007**, *129*, 10058–10059.

40. Zhan, Z.-Y. J.; Lynn, D. G. Chemical amplification through template-directed synthesis. *J. Am. Chem. Soc.* **1997**, *119*, 12420–12421.

41. Goodwin, J. T.; Lynn, D. G. Template-directed synthesis: Use of a reversible reaction. *J. Am. Chem. Soc.* **1992**, *114*, 9197–9198.

42. Leal, N. A.; Sukeda, M.; Benner, S. A. Dynamic assembly of primers on nucleic acid templates. *Nucleic Acids Res.* **2006**, *34*, 4702–4710.

43. Seeman, N. C. An overview of structural DNA nanotechnology. *Mol. Biotechnol.* **2007**, *37*, 246–257.

44. Aldaye, F. A.; Sleiman, H. F. Guest-mediated access to a single DNA nanostructure from a library of multiple assemblies. *J. Am. Chem. Soc.* **2007**, *129*, 10070–10071.

45. Krishnan-Ghosh, Y.; Whitney, A. M.; Balasubramanian, S. Dynamic covalent chemistry of self-templating PNA oligomers: Formation of a bimolecular PNA quadruplex. *Chem. Commun.* **2005**, 3068–3070.

46. Voshell, S. M.; Lee, S. J.; Gagné, M. R. The discovery of an enantioselective receptor for (–)-adenosine from a racemic dynamic combinatorial library. *J. Am. Chem. Soc.* **2006**, *128*, 12422–12423.

47. Bardají, E. G.; Freisinger, E.; Costisella, B.; Schalley, C. A.; Brüning, W.; Sabat, M.; Lippert, B. Mixed-metal (platinum, palladium), mixed-pyrimidine (uracil, cytosine) self-assembling metallacalix[n]arenes: Dynamic combinatorial chemistry with nucleobases and metal species. *Chem. Eur. J.* **2007**, *13*, 6019–6039.

Chapter 4

Complex Self-Sorting Systems

Soumyadip Ghosh and Lyle Isaacs

4.1. Introduction and Background

Beginning with the pioneering work of Cram, Lehn, and Pederson that defined the area of supramolecular chemistry as a contemporary discipline, the chemical community has focused significant attention on elucidating the fundamental aspects of noncovalent interactions between molecules [1–3]. For example, the use of H-bonds, π–π, and metal–ligand interactions as the driving force for the buildup of complex structures under thermodynamic control is now relatively well developed [4–9]. Accordingly, in the past decade the emphasis in supramolecular chemistry has shifted toward the development of self-assembled systems whose function derives from the precise orientation of the components relative to one another. For example, chemists have developed molecular machines that rely on intra-aggregate movements, chemical sensors that function by changes in the UV/Vis or fluorescence output of a chromophore upon aggregation, and membrane transporters that shuttle ions and molecules across the hydrophobic biological interface [10–13]. All of these functional systems rely on the design and a priori synthesis of specific molecules (e.g., hosts, chromophores, machines, and transporters) with specific structural features for specific functions.

4.1.1. Dynamic Combinatorial Chemistry

In the mid-1990s an alternative approach to the generation of functional supramolecular systems, dynamic combinatorial chemistry (DCC), was conceived and developed by several groups, with the earliest publications coming from the Sanders, Lehn, and Miller laboratories [14–16]. In DCC (Scheme 4.1), a series of building blocks (e.g., A_1–A_n, B_1–B_n) are mixed together and allowed to react with one another by the formation of *reversible* covalent bonds or noncovalent interactions. The set of compounds thus generated constitutes a *dynamic combinatorial library* (DCL), and the relative concentrations of the members of the DCL are governed entirely by thermodynamic considerations (e.g., concentrations and equilibrium constants). Next, a templating molecule (G) is added to the DCL, which induces a change in the equilibrium concentration of the various members of the DCL due to the formation of noncovalent complexes between the members of the DCL and the template as the new thermodynamic equilibrium is established. Accordingly, the concentration of certain members of the DCL becomes diminished upon addition of the template and certain members become enhanced. Usually, although not always [17–20], the members of the DCL whose concentrations become enhanced are those that bind most tightly (e.g., largest K_{eq}) to the template. The identification and separation or de novo synthesis of these tight binding receptors (hosts) then provides a route toward function (e.g., sensing, catalysis, and transport). The reader is referred to the other chapters of this book that focus more exclusively on DCC. In this chapter, we focus on a research area known as self-sorting, which we view as being intimately connected to the topic of DCC. Self-sorting is illustrated here with examples drawn from the research conducted in our laboratories at the University of Maryland, sprinkled with those that served as scientific stimulation for us and for subsequent developments in the field.

4.1.2. Self-Sorting

Self-sorting refers to the ability of a molecule to efficiently distinguish between self and nonself even within a complex mixture. As such the concept of self-sorting is intimately tied to the concepts of binding affinity, binding selectivity, and multicomponent mixtures that are commonly encountered in biological systems. A 1993 report from the Lehn group triggered our recognition of and interest in self-recognition (self-sorting) as a distinct research area (Scheme 4.2) [21]. In that report Lehn described the synthesis of a series of bipyridine oligomers (**1–4**) that were *individually* known to undergo well-defined self-assembly in the presence of Cu^+ by metal–ligand noncovalent interactions to afford the double helices ($\mathbf{1}_2 \cdot Cu_2^+ - \mathbf{4}_2 \cdot Cu_5^+$).

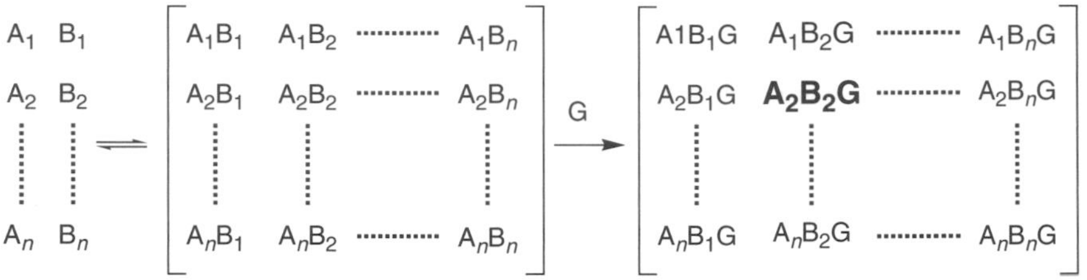

$$
\begin{array}{cc}
A_1 & B_1 \\
A_2 & B_2 \\
\vdots & \vdots \\
A_n & B_n
\end{array}
\rightleftharpoons
\begin{bmatrix}
A_1B_1 & A_1B_2 & \cdots & A_1B_n \\
A_2B_1 & A_2B_2 & \cdots & A_2B_n \\
\vdots & \vdots & & \vdots \\
A_nB_1 & A_nB_2 & \cdots & A_nB_n
\end{bmatrix}
\xrightarrow{\;G\;}
\begin{bmatrix}
A_1B_1G & A_1B_2G & \cdots & A_1B_nG \\
A_2B_1G & \mathbf{A_2B_2G} & \cdots & A_2B_nG \\
\vdots & \vdots & & \vdots \\
A_nB_1G & A_nB_2G & \cdots & A_nB_nG
\end{bmatrix}
$$

Scheme 4.1 Schematic representation of the equilibria involved in the generation of a DCL and the selection of a specific member of that DCL by application of a template. The relative concentrations of the guest-bound species are indicated by their font size.

Lehn then asked the simple but far-reaching question of what would happen if all four oligomers were mixed together simultaneously. Would the information encoded within the molecular structure of **1–4** under the readout of Cu^+ result in the formation of a simple mixture of double helicates (e.g., $\mathbf{1_2} \cdot Cu_2^+ - \mathbf{4_2} \cdot Cu_5^+$) by a self-recognition (self-sorting) process based on oligomer length, or would crossover aggregation between ligands of different length occur? In the experiment a mixture comprising ($\mathbf{1_2} \cdot Cu_2^+ - \mathbf{4_2} \cdot Cu_5^+$) was observed. In a related experiment Lehn used the coordination number preferences of Cu^+ and Ni^{2+} to direct the self-recognition (self-sorting) of two different bipyridine trimers [21]. In this manner Lehn showed that some complex mixtures can undergo surprisingly simple behavior by application of metal–ligand *noncovalent interactions*.

The first report that we are aware of in the chemical literature to use the term "self-sorting" comes from Sanders' group [22]. In this paper, Sanders reports the thermodynamically controlled transesterification reaction ($KOCH_3$, 18-crown-6, toluene, reflux) of a mixture of predisposed building blocks **5** and **6** (Scheme 4.3). Remarkably, rather than a complex mixture of macrocycles comprising both building blocks, a simple mixture of dimeric macrocycle **7** derived from **6** and trimeric macrocycle **8** derived from **5** was observed. This paper was very significant in that it began to define self-sorting as an area of research and was the first example of a self-sorting system based on reversible covalent bond formation under thermodynamic control within a mixture.

4.2. Entry into Self-Sorting Research at the University of Maryland

As a newly formed research group at the University of Maryland in 1998 we were interested in starting research projects in two areas: (1) DCC

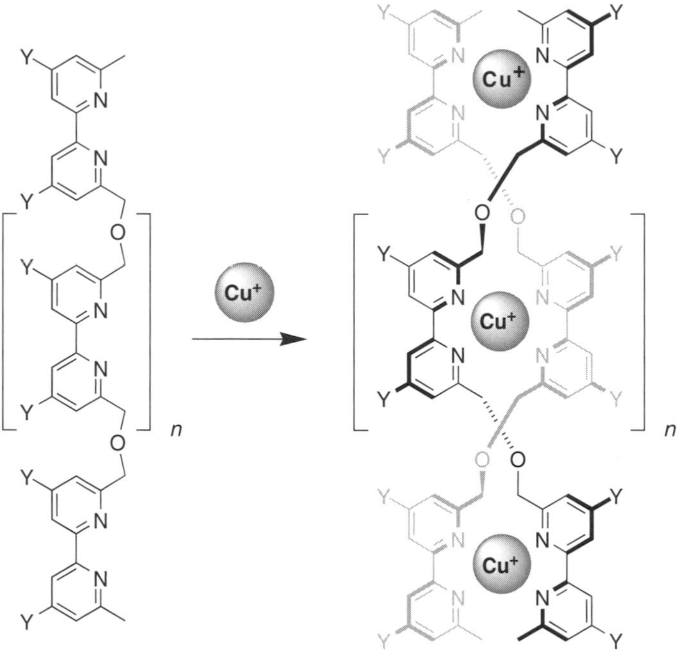

$$n = 0 \quad \mathbf{1}$$
$$n = 1 \quad \mathbf{2}$$
$$n = 2 \quad \mathbf{3}$$
$$n = 3 \quad \mathbf{4}$$

$Y = H, CONEt_3, CO_2Et$

$n = 0 \quad [Cu_2(\mathbf{1})_2]^{2+}$
$n = 1 \quad [Cu_3(\mathbf{2})_2]^{3+}$
$n = 2 \quad [Cu_4(\mathbf{3})_2]^{4+}$
$n = 3 \quad [Cu_5(\mathbf{4})_2]^{5+}$

Scheme 4.2 Self-recognition within mixtures of bipyridine-based Cu^+ double helicates.

on self-assembled monolayers and on the side chains of suitable derivatized polymer backbones [23] and (2) the cucurbit[n]uril family of macrocycles [24,25]. Although we were never able to get the DCC project rolling,
the ideas continued to percolate in my head for several years and finally
re-emerged in the form of self-sorting in 2002. A key stimulus for our
work in the self-sorting area came from the group of our colleague, Professor Jeffery Davis, who was studying the self-association of isoguanosine and guanosine derivatives **9** and **10** (Scheme 4.4). They made several
remarkable findings: (1) racemic guanosine **10** undergoes enantiomeric
self-recognition to form hexadecameric $\mathbf{10}_{16} \cdot 2Ba^{2+}$, (2) isoguanosine **9**
undergoes self-association to form decameric $\mathbf{9}_{10} \cdot Ba^{2+}$, and (3) mixtures
of **9** and **10** undergo self-sorting to form a mixture of $\mathbf{10}_{16} \cdot 2Ba^{2+}$ and
$\mathbf{9}_{10} \cdot Ba^{2+}$ without significant crossover aggregation processes [26–28].

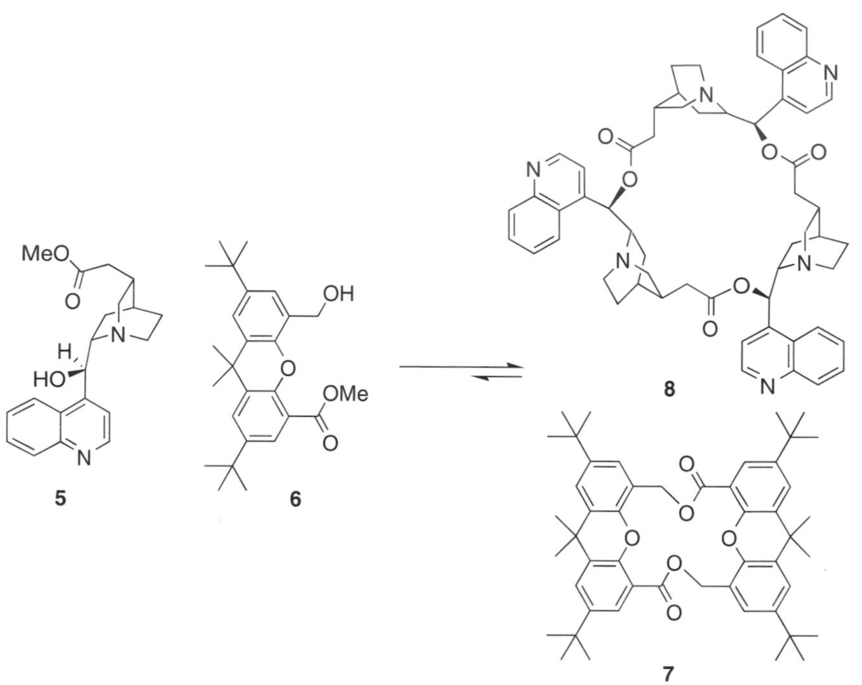

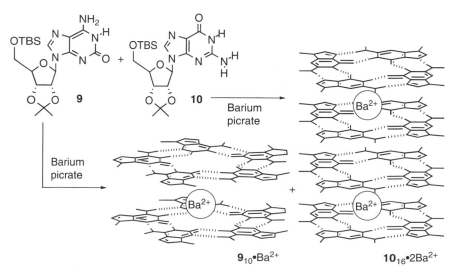

Scheme 4.3   Self-sorting based on thermodynamically controlled transesterification during macrocyclization of predisposed building blocks.

Scheme 4.4   Self-sorting of **9** and **10** in the presence of barium picrate leads to the formation of $10_{16} \cdot 2Ba^{2+}$ and $9_{10} \cdot Ba^{2+}$.

At around the same time, Dr. Dariusz Witt in our research group was learning how to synthesize the methylene-bridged glycoluril dimer substructure [29–31], which is the fundamental building block of the cucurbit[n]uril family of macrocycles that would allow us to pursue the second project described above. As an intermediate goal, we wanted to perform self-assembly in water using these C-shaped molecular clips [32] and decided to synthesize (±)-**11** with the hope that it would form a square-shaped aggregate by metal–ligand interactions [8,9] and the hydrophobic effect (Scheme 4.5). Experimentally, Darek found that (±)-**11** underwent an enantiomeric self-recognition process triggered by the addition of **12** during the formation of the racemic mixture (+)-**11**$_2$ · **12**$_2$ and (−)-**11**$_2$ · **12**$_2$ [33a]. The self-recognition of one enantiomer of a racemic ligand during self-assembly to form homochiral metal complexes can be viewed as a form of self-sorting (e.g., the ability to distinguish between a ligand and its enantiomer even within the more complex mixture of the two enantiomers).

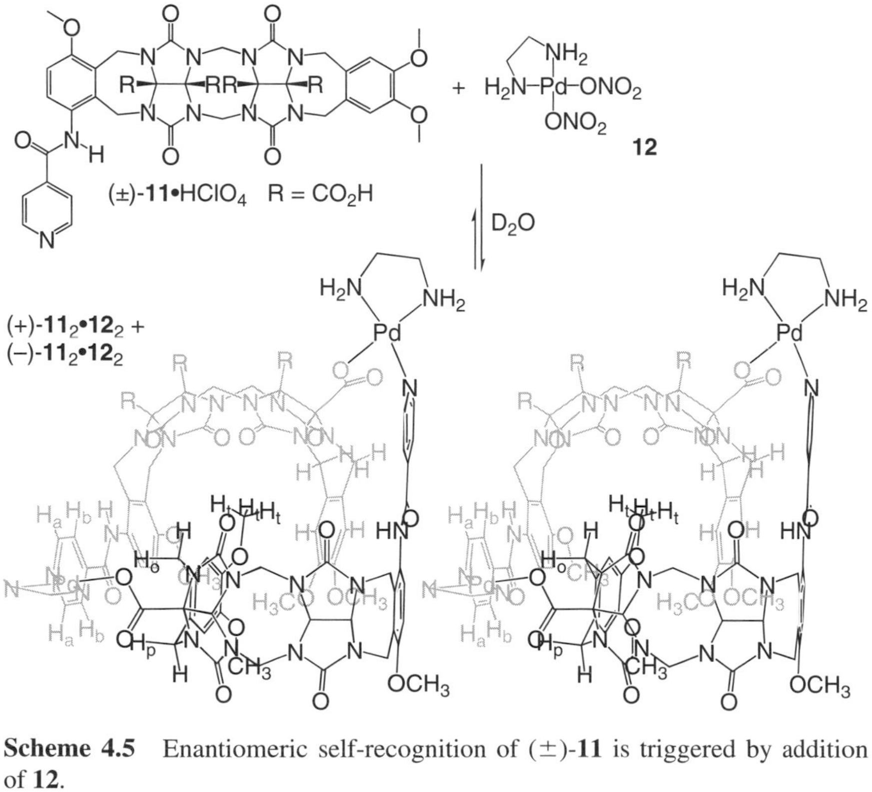

Scheme 4.5 Enantiomeric self-recognition of (±)-**11** is triggered by addition of **12**.

Although this interesting result can be recognized in hindsight as the starting point for our entry into the area of self-sorting research, our realization of the true power of self-sorting systems and their intimate connection to biology would have to wait. When Anxin Wu, now professor at Central China Normal University in Wuhan, joined the group in early 2001, we decided to follow up on this earlier research by preparing **13** and (±)-**14**, which contained two pyridyl groups [33b]. We anticipated that carboxylic acids **13a** and (±)-**14a** might form large hydrophobically driven aggregates when combined with **12** in water. Unfortunately, no well-defined aggregates were observed by ^{1}H NMR spectroscopy. Although this result was disappointing to us, Wu discovered that esters **13e** and (±)-**14e** undergo tight dimerization ($K_a \geq 10^6$ M^{-1}) in CDCl$_3$ solution to deliver **13e$_2$** and (+)-**14e** · (−)-**14e** by a heterochiral recognition process (Scheme 4.6). At first glance the formation of these simple H-bonded dimers did not appear to us to be the makings of a high impact publication. Upon further reflection, and based on the uncommon ability of **14e** to undergo high fidelity heterochiral recognition, we decided to formulate the paper in terms of self- versus nonself-recognition. Naturally, therefore, we decided to investigate the mixture comprising **13e** and (±)-**14e** in CDCl$_3$. ^{1}H NMR spectroscopy of this mixture was simply the superposition of the ^{1}H NMR spectra of its components, which is the spectroscopic fingerprint of a self-sorting process.

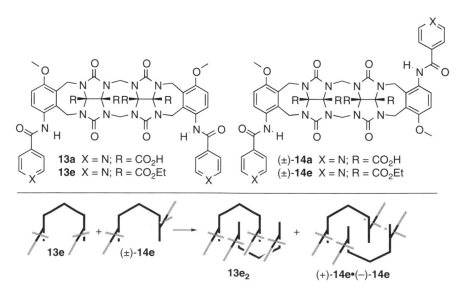

Scheme 4.6 Self-sorting behavior of C-shaped molecular clips **13** and (±)-**14**.

4.3. How is Self-Sorting Different from Self-Assembly?

Over the years many people have posed this question to us. To answer we first begin with our working definition of self-assembly. Self-assembly is the spontaneous high fidelity synthesis of a (note: singular) higher order structure from its components usually under thermodynamic control. Self-sorting refers to the spontaneous high fidelity synthesis of a system (note: plural) of higher order structures from their components usually under thermodynamic control. The critical distinction between self-assembly and self-sorting, therefore, is that self-sorting occurs within complex multicomponent mixtures, whereas self-assembly occurs from a single set of components. At this point the usual response is "So what?" To the extent that self-sorting systems are merely the superposition of a series of well-defined aggregates from the literature, we agree with this pessimistic assessment.

With proper design and implementation, however, it is possible to construct a self-sorting system whose behavior is different from its components [33c]. For example, consider the simple system comprising two hosts (A and B) and two guests (M and N) that can form four possible host–guest complexes (AM, AN, BM, and BN). We fix the total concentrations of hosts A and B ($[A_{tot}]$ and $[B_{tot}]$) at 1 mM and choose the four equilibrium constants such that host A (10^4-fold) and host B (10-fold) both prefer guest M (Scheme 4.7). The various mole fraction definitions (Scheme 4.7c) are used to construct a plot (Scheme 4.7d) of the composition of the mixture as a function of total guest concentration ($[M_{tot}] = [N_{tot}]$). When $[A_{tot}] = [B_{tot}] \geq [M_{tot}] = [N_{tot}]$, complexes AM and BN dominate because A binds M 100-fold more tightly than B binds M. The excess free energy obtained from forming AM can be used to force B to accept N despite its individual preference for M. The critical realization is that because *self-sorting systems minimize the overall free energy* of the entire system, unusual behavior that differs from the individual components may occur. The ultimate example of this type of behavior is embodied in living systems whose multitude of components undergo well-defined self-sorting processes that are orchestrated both in time and in space. No one would consider isolated proteins, nucleic acids, or lipids to be alive, but few can dispute the remarkable emergent behaviors that occur when these components are present as part of the cell. The ongoing goal of our research in the self-sorting area is to go beyond those systems that are simply equal to the sum of their parts and to move toward systems that exhibit behavior similar to those observed in nature (e.g., metastable energy dissipative rather than thermodynamically stable systems, compartmentation, and catalytic events to alter the free-energy landscape of complex mixtures).

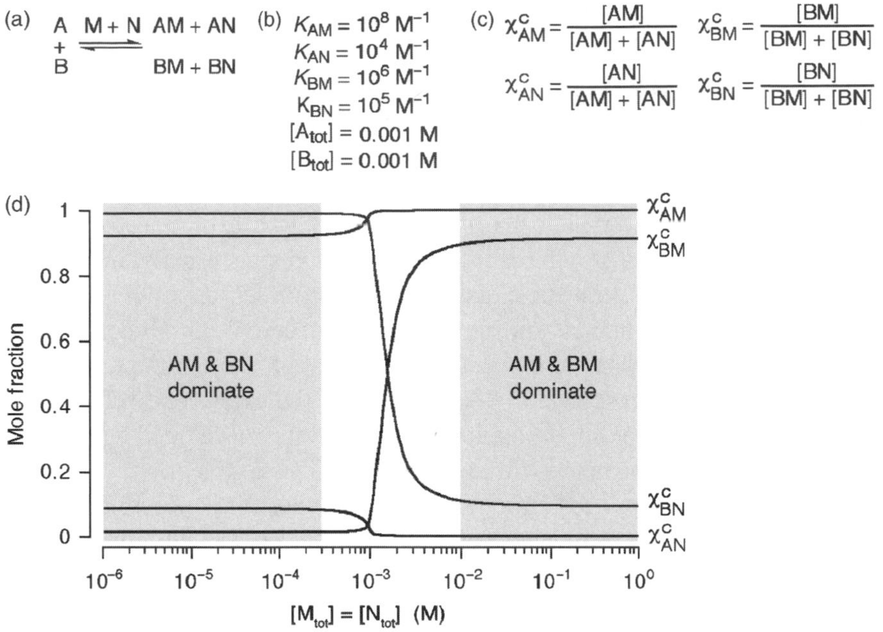

(a) A   M + N   AM + AN
    +   ⇌
    B   BM + BN

(b) $K_{AM} = 10^8 \text{ M}^{-1}$
$K_{AN} = 10^4 \text{ M}^{-1}$
$K_{BM} = 10^6 \text{ M}^{-1}$
$K_{BN} = 10^5 \text{ M}^{-1}$
$[A_{tot}] = 0.001 \text{ M}$
$[B_{tot}] = 0.001 \text{ M}$

(c) $\chi^c_{AM} = \dfrac{[AM]}{[AM] + [AN]}$ $\chi^c_{BM} = \dfrac{[BM]}{[BM] + [BN]}$

$\chi^c_{AN} = \dfrac{[AN]}{[AM] + [AN]}$ $\chi^c_{BN} = \dfrac{[BN]}{[BM] + [BN]}$

(d)

Scheme 4.7 Stoichiometry-induced partner displacement in a four-component mixture: (**a**) equilibria considered, (**b**) constraints imposed, (**c**) mole fraction definitions, and (**d**) a plot of mole fraction versus guest concentration ($[M_{tot}] = [N_{tot}]$).

4.4. Development of Complex Self-Sorting Systems at the University of Maryland

This section describes the development of self-sorting systems that has occurred in our laboratory over the past 5 years [34].

4.4.1. Self-Sorting: The Exception or the Rule

The self-sorting systems described above were all based on the use of sets of compounds that possessed very similar structures. Accordingly, it was viewed by the community as quite exceptional behavior that these systems undergo self-sorting based on metal–ligand, π–π, and H-bonding interactions. In fact, despite the wide range of remarkably complex and functional systems that had been self-assembled over the years, there was a perception that synthetic hosts and self-assembled systems were inferior to natural systems in terms of strength and selectivity of binding. Based on the studies described above we came to the idea that the widespread perception of the community might be incorrect. We decided, therefore, to create a multicomponent system comprising the molecular building blocks of a series of well-defined aggregates from the literature and asked whether those compounds possessed the ability to efficiently distinguish between self and nonself even within a complex mixture. Would the system undergo a high fidelity self-sorting process or would crossover heteromeric aggregation occur? For this purpose we selected **9** and **10** described above and added two new molecular clips **15** and **16**, the components of Meijer's ureidopyrimidinone dimer **17**$_2$ [35], Rebek's tennis ball **18**$_2$ and calixarene tetraurea capsule **19**$_2$ [36, 37], and **20** and **21**, which are the components of Reinhoudt's double rosette (Fig. 4.1) [38]. We next measured the ^{1}H NMR spectrum of each aggregate separately in CDCl$_3$ solution (Fig. 4.2a–h) and then measured the ^{1}H NMR spectrum of the eight-component mixture (Fig. 4.2i) [34a]. Remarkably, the ^{1}H NMR spectrum of the eight-component mixture is simply equal to the sum of the ^{1}H NMR spectra of its component aggregates. This spectroscopic earmark indicates that this nine-component mixture undergoes a high fidelity self-sorting process. We conclude that the precise pattern of H-bond donors and acceptors, the spatial distribution of those H-bonding groups, and the presence of closed networks of H-bonds are factors that favor self-sorting rather than crossover heteromeric aggregation. We also studied the influence of several key variables on the self-sorting process—temperature, concentration, values of K_{eq}, and the presence of H-bonding competitors—by a combination of simulation and experiment. Although it is clear that many systems do undergo heteromeric aggregation rather than self-sorting, this study led us to conclude that the scope of the systems that are sufficiently selective to undergo self-sorting processes is much wider than previously appreciated.

4.4.2. Social Self-Sorting in Aqueous Solution

Although the nine-component self-sorting system described above increased the complexity of designed synthetic self-sorting systems beyond what was possible previously, it suffered a number of drawbacks. First, this self-sorting process occurred in CDCl$_3$ solution and was driven exclusively by H-bonds; in contrast, nature's self-sorting systems occur in aqueous solution driven by myriad noncovalent interactions. Second, the above system was constructed using mainly self-association processes, a subset of self-sorting processes dubbed *narcissistic self-sorting* by Anderson in his lovely work on oligomeric porphyrin ladders [39]. Narcissistic self-sorting systems are particularly limited because strong self-association limits the number of different partners a molecule may have over its

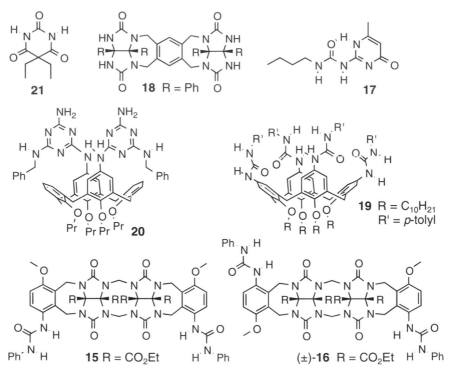

Figure 4.1 Chemical structures of compounds used in H-bond directed self-sorting in CDCl₃.

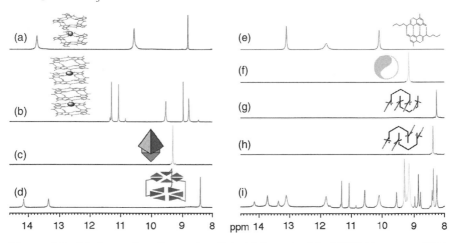

Figure 4.2 Hydrogen bonding region (8.0–14.5 ppm) of the ¹H NMR spectra (H₂O sat. CDCl₃, 500 MHz, 298 K) recorded for (a) **9₁₀** · Ba²⁺ + 2Pic⁻, (b) **10₁₆** · 2Ba²⁺ 4Pic⁻, (c) **19₂**, (d) **20₃** · **21₆**, (e) **17₂**, (f) **18₂**, (g) **15₂**, (h) (+)-**16** · (−)-**16**, (i) a self-sorted mixture comprising **9₁₀** · Ba²⁺ + 2Pic⁻, **10₁₆** · 2Ba²⁺ 4Pic⁻, **19₂**, **20₃** · **21₆**, **17₂**, **18₂**, **15₂**, and (+)-**16** · (−)-**16**. The representations depict the species present in solution. The resonances are color coded to aid comparison. See insert for color representation of this figure.

lifetime to one. In contrast, it is possible to imagine a self-sorting system composed of host–guest pairs. The advantage of such a host–guest-based self-sorting system is that it is potentially environmentally responsive in that the addition of a better binding guest to the mixture would result in a change in composition of the self-sorting system. We refer to a host–guest-based self-sorting system as a *social self-sorting* system. With these considerations in mind we selected compounds (±)-**11**, **12**, and **22–28** as the components of a social self-sorting system in aqueous solution (Fig. 4.3) [33c]. Compound (±)-**11** is known to undergo enantiomeric self-recognition triggered by **12**, molecular clip **23** undergoes tight dimerization, and cryptand **22** binds tightly to K$^+$ ion in water. The molecular containers β-CD (**26**), CB[6], and CB[8] form discrete host–guest complexes with adamantane carboxylic acid (**28**), hexanediammonium ion (**24**), and the charge transfer complex comprising dihydroxynaphthalene and methyl viologen (**25 · 27**), respectively. Once again we use ^{1}H NMR as our analytical technique and measure the spectrum of each of the host–guest pairs separately and then measure the spectrum of the 12-component mixture. The ^{1}H NMR spectrum of the mixture is simply the sum of the ^{1}H NMR spectra of its components, which indicates this system undergoes a high fidelity social self-sorting process. We also studied the influence of pH, temperature, equilibrium constant, and host:guest stoichiometry on the fidelity of self-sorting by a combination of simulation and experiment. The significance of this study lies in the demonstration that less directional interactions like ion–dipole and π–π interactions and the hydrophobic effect can be used to drive self-sorting in water in much the same way

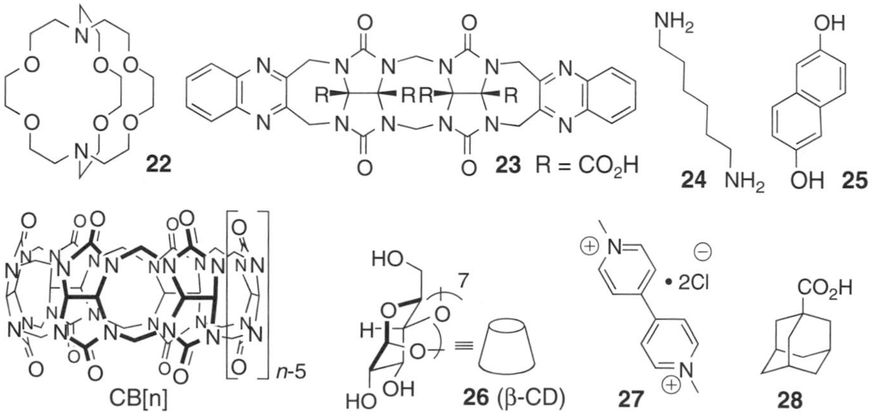

Figure 4.3 Chemical structures of compounds used for social self-sorting in aqueous solution driven by ion–dipole, metal–ligand, π–π, and hydrophobic interactions.

more directional H-bonds can be used in $CDCl_3$ solution. In addition, the use of host–guest pairs as the basis of a self-sorting system (social self-sorting) offers the potential for environmental responsiveness in the form of tighter binding competitive guests. We expect this responsiveness will be one of the vehicles to achieve biomimetic function.

4.4.3. The CB[n] Family of Macrocycles are Prime Components for the Construction of Self-Sorting Systems

In the course of preparing the 12-component social self-sorting system described above and in unpublished investigations of the complexation between CB[n] hosts and their guests, we were surprised that well-defined complexes were obtained in such high fidelity processes. This result suggested to us that the high binding affinity (K_a up to 10^7 M^{-1}) and selectivity delineated by the pioneering work of Mock for CB[6] would also be observed *individually* for the larger CB[n] homologues (CB[7] and CB[8]) and also that *collectively* the selectivity of CB[6], CB[7], and CB[8] toward a common guest might be large. Accordingly, we measured the binding affinity of CB[6], CB[7], and CB[8] toward a series of guests (**29–36**) by ^{1}H NMR competition experiments [34b]. A selection of these values of K_a is presented in Table 4.1. Remarkably, the range of values of K_a spans more than 10 orders of magnitude! Several series of complexes deserve comment (Scheme 4.8). For example, consider the CB[7] · **29** and CB[8] · **29** complexes that differ in stability by over seven orders of magnitude. Apparently, **29** is slightly too large for CB[7], but has a good size and shape match with the cavity of CB[8]. Similarly, **35** prefers to bind to CB[7] over CB[6] due

Table 4. 1   Values of K_a (M^{-1}) measured for the binding between CB[n] hosts and guests (D_2O, pD 4.74, 25°C)

	CB[6]	CB[7]	CB[8]
29	–	2.5×10^4	4.3×10^{11}
30	nb	1.5×10^5	–
31	8980	8.4×10^6	–
32	–	1.8×10^7	5.8×10^{10}
33	nb	1.8×10^7	nb
24	4.5×10^8	9.0×10^7	–
34	nb	8.9×10^8	nb
35	550	1.8×10^9	–
36	–	4.2×10^{12}	8.2×10^8

Note: – indicates not measured; nb indicates no binding.

to size considerations. There are, however, situations where a guest binds tighter to the smaller CB[n] homologue. For example, **24** prefers to bind to CB[6] relative to CB[7] (5-fold), and **36** prefers to bind to CB[7] relative to CB[8] (5000-fold); we attribute these preferences to a better size match between the guest and the smaller CB[n] homologue. A final intriguing entry in Table 4.1 concerns the CB[7] · **32** and CB[8] · **32** complexes that differ in affinity by over 3000-fold. In the CB[7] · **32** complex only one arm of **32** can fit inside CB[7]; in contrast CB[8] induces a U-shaped turn of guest **32**, which better fills the cavity of CB[8]. This result is significant because it suggested to us that CB[8] is capable of controlling the folding of abiotic oligomers in water (*vide infra*).

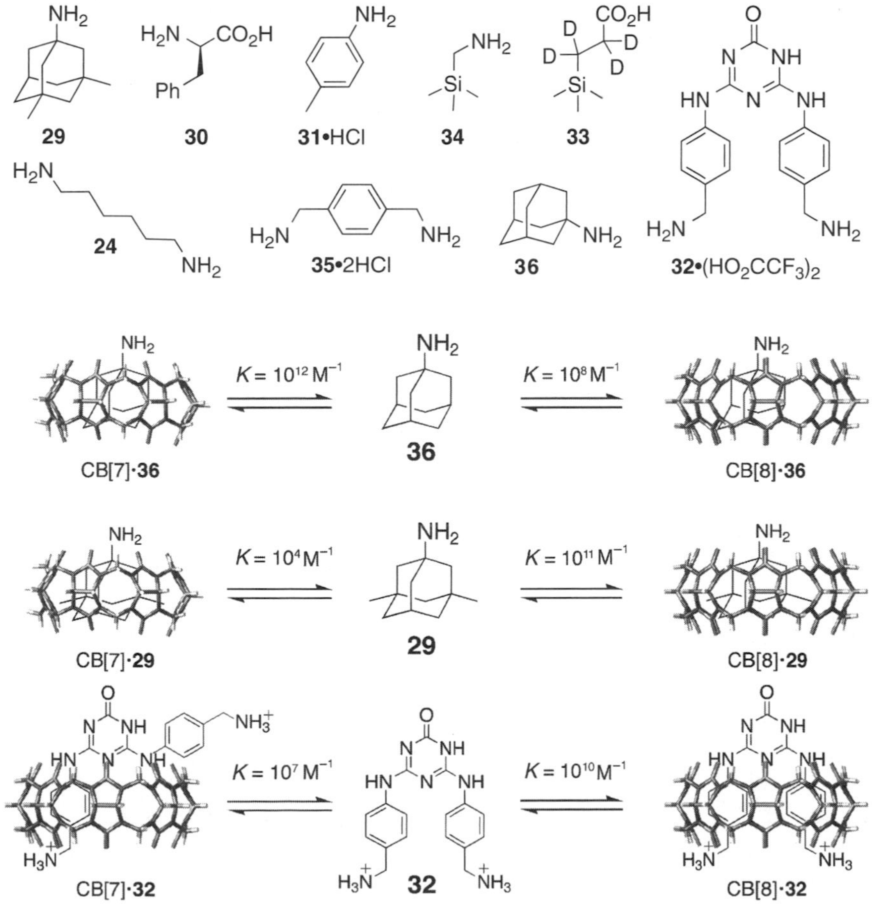

Scheme 4.8 Illustration of the competition between a single guest for different-sized CB[n].

4.4.4. Kinetic Self-Sorting

The sections above describe the preparation of self-sorting systems under thermodynamic control. In the design of such systems the only variables that are relevant are the concentrations of the components and the equilibrium constants of the various complexes. Living systems, of course, do not reach equilibrium until they are dead and instead use strategies to maintain metastable states and otherwise control the approach toward equilibrium. One of the common strategies employed by natural processes is the use of molecular species containing multiple binding epitopes, each of which serves a specific functional role. To introduce elements of temporal control into our self-sorting systems, we decided, therefore, to synthesize and study guests containing two distinct binding epitopes (**37–39**), which we refer to as two-faced guests (Scheme 4.9) [34c]. In designing these guests we planned to take advantage of not only the well-defined thermodynamic preferences of CB[*n*] toward suitable cationic guests, but also the fact that their kinetics of association and dissociation are known to span many orders of magnitude [40–43]. Experimentally, when a solution of CB[6] and CB[7] was mixed with a solution of **37** and **40**, we observed the initial formation of the CB[6] · **37** and CB[7] · **40** complexes. After a period of 56 days this kinetic preference was lost and a thermodynamic self-sorting system comprising CB[6] · **40** and CB[7] · **37** was obtained! Experiments involving two-faced guests with longer (e.g., **38** and **39**) alkylammonium tails as CB[6] binding epitopes compromise the fidelity of the kinetic self-sorting state. By a combination of experiment and simulation we were able to determine that the major factor controlling the high fidelity of kinetic self-sorting is the fact that **40** associates faster with CB[7] than **37** does despite the fact that the CB[7] · **37** complex is thermodynamically more stable than CB[7] · **40**. This situation probably arises because the ureidyl C=O lined portals of CB[*n*] are narrower than the cavity they guard, which may result in large barriers to association and dissociation for guests optimally filling the CB[*n*] cavity [40–43]. The influence of metal cations (identity and concentration) on the fidelity of the kinetic and thermodynamic self-sorting states was also investigated. One of the most interesting aspects of this investigation was the *post facto* deconstruction of the system, which revealed that the CB[6] · **40** complex is governed by remarkably slow kinetics of association (k_{in} = 0.0012 M^{-1} s^{-1}) and dissociation (k_{out} = 8.5 × 10^{-10} s^{-1}). This dissociation rate constant, which is approximately 100-fold slower than biotin · avidin, corresponds to a half-life of 26 years at room temperature! Because the outcome of self-sorting experiments simultaneously probes the entire matrix of kinetic constants

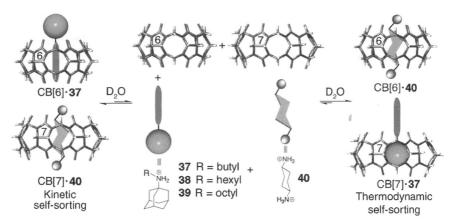

Scheme 4.9 Kinetic and thermodynamic self-sorting based on guests with two binding epitopes.

and thermodynamic parameters (Scheme 4.9), these simple experiments are capable of revealing truly anomalous host–guest binding experiments rapidly. As uncovering and understanding such anomalous binding events is one of the key subjects for supramolecular chemists, we believe that self-sorting systems have much to offer to the community.

4.4.5. Self-Sorting Processes Control the Folding, Forced Unfolding, and Refolding of an Abiotic Oligomer in Water

Based on the high association constants observed for CB[7] · **32** and CB[8] · **32**, the high selectivity observed between these two complexes, and the interesting folding induced by CB[8], we wondered whether longer oligomers might be induced to exhibit well-defined conformational preferences in the presence of various CB[n] (n = 7, 8, 10). For this purpose, we designed and synthesized a number of arylene–triazene oligomers exemplified by compound **41**. Compound **41** contains four (guanidine-like) triazene-N bonds (highlighted with curved arrows), which may adopt two different rotamers (Fig. 4.4). Of the 2^4 (16) possible conformations, 10 are unique. Through a combination of NMR spectroscopy and X-ray crystallography we were able to determine that the complexation of **41** with CB[7], CB[8], or CB[10] leads to the formation of well-defined CB[7] · (a,s,s,a)-**41** · CB[7], CB[8] · (a,a,a,s)-**41**, and CB[10] · (a,a,a,a)-**41** complexes (Scheme 4.10) [34d]. Perhaps most interesting, however, is how the high affinity and selectivity of CB[n] complexes can be used as a thermodynamic driving force to sequentially induce the folding, unfolding, and refolding of abiotic oligomer

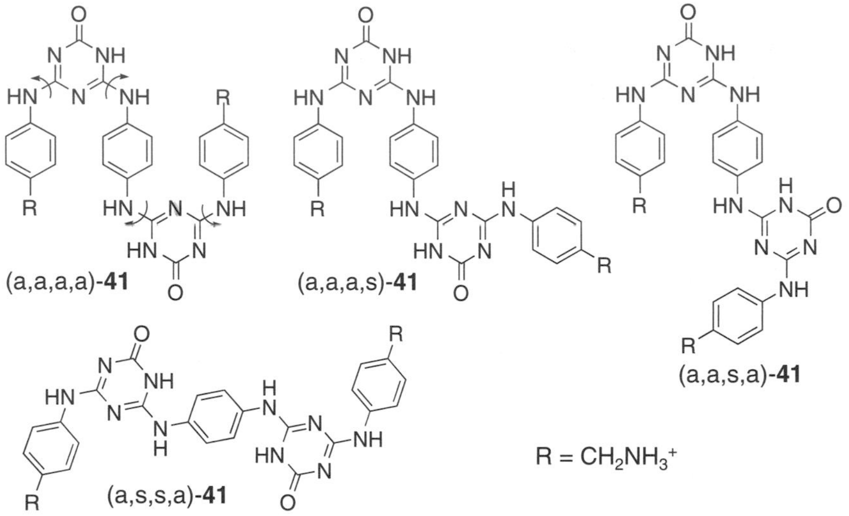

Figure 4.4 Chemical structures of 4 of the 10 conformational isomers of **41**. Arrows highlight the four twofold rotors.

41 into four different conformations. We first add CB[8] to a solution of **41**, which populates the full 10-component conformational ensemble open to **41**; this induces the folding of **41** to yield the CB[8] · (a,a,a,s)-**41** conformation by maximization of ion–dipole and H-bonding to the ureidyl C=O lined portals of CB[8] and the hydrophobic effect. By addition of **29** (1 equiv.) and CB[7] (2 equiv.) it is possible to eject **41** from the cavity of CB[8] under the formation of CB[8] · **29** and then refold **41** with the help of CB[7] to yield CB[7] · (a,s,s,a)-**41** · CB[7]. In this process the high affinity of **29** for CB[8] ($K_a = 4.3 \times 10^{11}$ M^{-1}) ensures the transformation into a high fidelity self-sorting state. If this transformation is conducted stepwise by the addition of CB[7] first, it is possible to spectroscopically observe the intermediacy of CB[8] · (a,a,s,a)-**41** · CB[7]. Remarkably, control experiments show that the binding of CB[7] to the tail of guest **41** *catalyzes* its dissociation from the cavity of CB[8]. Finally, addition of **36** (2 equiv.) to CB[7] · (a,s,s,a)-**41** · CB[7] results in the formation of CB[7] · **36** ($K_a = 4.2 \times 10^{12}$ M^{-1}) and ejection of **41** into free solution where it forms its equilibrium mixture of conformers. Subsequently, **41** ejects CB[5] from the CB[10] · CB[5] complex to yield a self-sorted state comprising CB[10] · (a,a,a,a)-**41**, free CB[5], CB[8] · **29**, and CB[7] · **36**. This study illustrates how it is possible to use high affinity and highly selective receptors like CB[n] and their associated large values of ΔG of binding to mimic an important biological event, namely, the folding, unfolding, and refolding of an oligomeric compound in water.

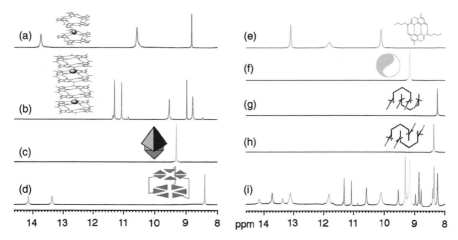

Figure 4.2 Hydrogen bonding region (8.0–14.5 ppm) of the ^{1}H NMR spectra (H$_2$O sat. CDCl$_3$, 500 MHz, 298 K) recorded for (**a**) **9**$_{10}$ · Ba^{2+} + 2Pic$^-$, (**b**) **10**$_{16}$ · 2Ba^{2+} 4Pic$^-$, (**c**) **19**$_2$, (**d**) **20**$_3$ · **21**$_6$, (**e**) **17**$_2$, (**f**) **18**$_2$, (**g**) **15**$_2$, (**h**) (+)-**16** · (−)-**16**, (**i**) a self-sorted mixture comprising **9**$_{10}$ · Ba^{2+} + 2Pic$^-$, **10**$_{16}$ · 2Ba^{2+} 4Pic$^-$, **19**$_2$, **20**$_3$ · **21**$_6$, **17**$_2$, **18**$_2$, **15**$_2$, and (+)-**16** · (−)-**16**. The representations depict the species present in solution. The resonances are color coded to aid comparison. See pages 127–128 for text discussion of this figure.

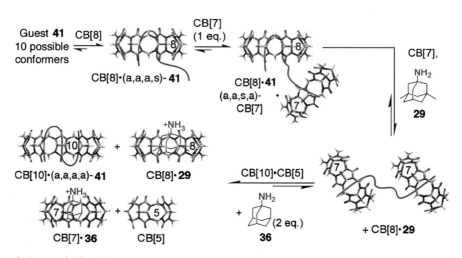

Scheme 4.10 The sequential addition of various CB[*n*] and guests to **41** induces folding, forced unfolding, and refolding of **41** into four different conformations. See pages 133–135 for text discussion of this figure.

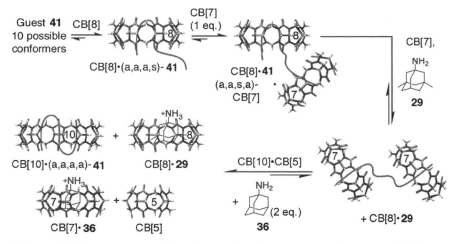

Scheme 4.10 The sequential addition of various CB[n] and guests to **41** induces folding, forced unfolding, and refolding of **41** into four different conformations. See insert for color representation of this figure.

4.5. Selected Examples of Self-Sorting Systems from Other Laboratories

A number of laboratories have been pursuing the development of self-sorting systems either explicitly or implicitly. In this section we present a selection of those systems based on a variety of noncovalent interactions.

4.5.1. Self-Sorting Systems Based on Porphyrin Coordination Chemistry

The earliest example of a porphyrin-derived self-sorting system comes from Taylor and Anderson [39]. Anderson's group studied the formation of porphyrin ladders from butadiyne-linked linear porphyrin dimer through hexamer in combination with diazabicyclooctane. They observed a variety of interesting behavior including positive cooperativity, two-state assembly, large Hill coefficients, and narcissistic self-sorting. More recently, Osuka's group studied the assembly and self-sorting of *meso–meso*-linked porphyrin dimers bearing pyridyl substituents [44–46]. A beautiful example arises from the assembly of *meso*-cinchomeronimide-appended diporphyrin **42** [45]. Compound **42** contains three axes of chirality—the *meso–meso* porphyrin link and the two cinchomeronimide–porphyrin linkages—which lead to six different atropisomers enumerated in Scheme 4.11. These six atropisomers exist as three pairs of enantiomers, namely, the in–in, in–out,

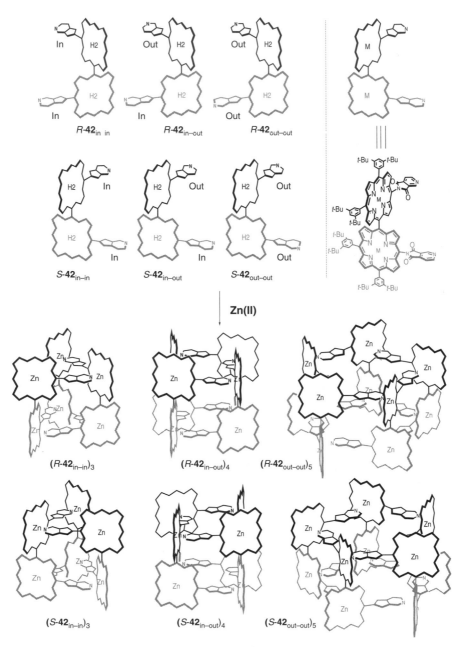

Scheme 4.11 Homochiral self-sorting assembly of *meso*-linked porphyrin **42**.

and out–out atropisomers that differ in the angle between the two coordinating N-atoms of the cinchomeronimide groups. Remarkably, the atropisomer with the smallest angle 42_{in-in} leads to a triangular trimeric porphyrin box $(42_{in-in})_3$, whereas the atropisomers with larger angles 42_{in-out} and $42_{out-out}$ lead to tetrameric and pentameric porphyrin boxes $(42_{in-out})_4$ and $(42_{out-out})_5$, respectively. Even more remarkable is that this self-sorting assembly process occurs in a homochiral fashion whereby all molecules of **42** in a given assembly are of a single handedness [e.g., $(R\text{-}42_{in-in})_3$ and $(S\text{-}42_{in-in})_3$]. This result nicely demonstrates that subtle conformational and stereochemical information can be efficiently translated from simple building blocks to far more complex assemblies.

4.5.2. Self-Sorting Based on Hydrogen Bonding Interactions

An early example of a self-sorting system based on H-bonding interactions was reported by Reinhoudt and coworkers [38]. They found that when calixarene dimelamine (**20**), a dicalixarene tetramelamine, and diethylbarbituric acid (**21**) were mixed in $CDCl_3$, a high fidelity self-sorting process occurred that delivered bis-rosette $20_3 \cdot 21_6$ and the corresponding tetrarosette. Zimmerman's group has studied the self-assembly of H-bonding modules derivatized with dendrons of different generations in nonpolar solvents by H-bonding interactions [47, 48]. For example, Zimmerman and coworkers found that **43**, which contains donor-acceptor-acceptor (DAA) and donor-donor-acceptor (DDA) H-bonding faces with 60° relative orientation, undergoes self-assembly to yield hexamer 43_6 (Scheme 4.12) as evidenced by size exclusion chromatography (SEC). Similarly, dimeric compound **44** with its self-complementary DDAA faces forms an exceptionally stable cyclic hexamer 44_6 by H-bonding interactions. Very interestingly, a mixture of **43** and **44** undergoes a high fidelity self-sorting process to yield the individual hexamers 43_6 and 44_6 driven by the different patterns of H-bonds and relative orientation of their faces; SEC monitoring of this mixture over the course of 53 days does not reveal any evidence of crossover heteromeric aggregation. Quite interestingly, when a mixture of **43-G1** and **43-G3**, which contain identical H-bonding information but different-sized dendritic wedges on their periphery, is prepared, a high fidelity social self-sorting process is observed under the formation of the mixed aggregate $(\mathbf{43\text{-}G1})_3 \cdot (\mathbf{43\text{-}G3})_3$ wherein small dendritic wedges are adjacent to large dendritic wedges, and vice versa. This process proceeds via the kinetic intermediacy of a DCL of the 11 distinct hexameric aggregates containing all combinations of 1–5 equiv. of **43-G1** and **43-G3**.

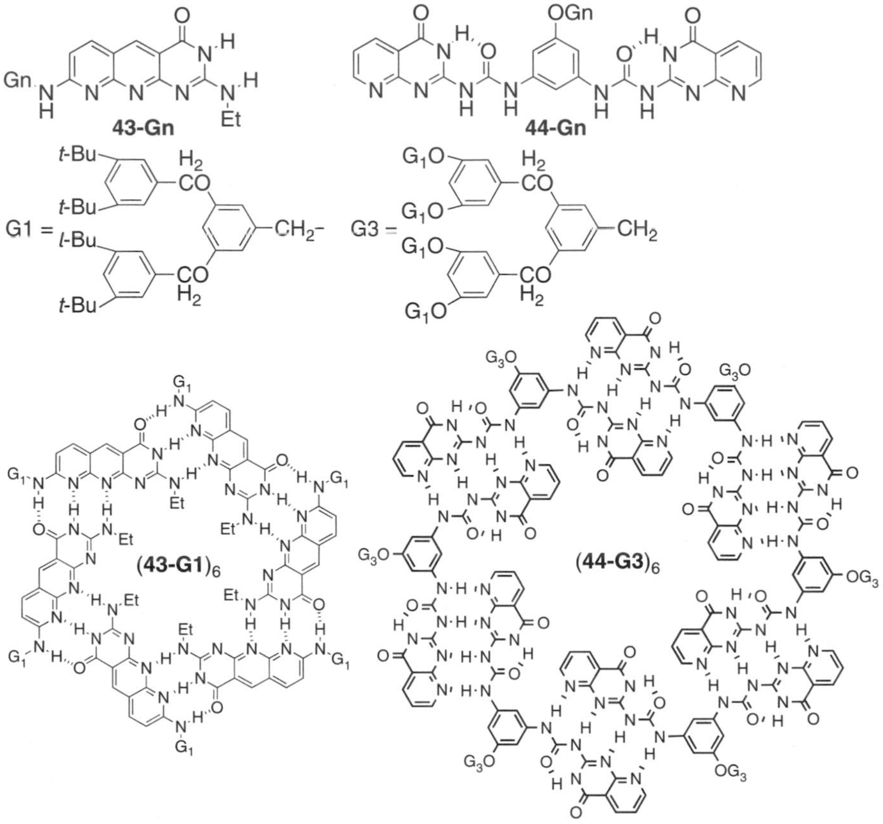

Scheme 4.12 Hexameric assemblies based on DDA–AAD H-bonding motif.

4.5.3. Self-Sorting Coiled Coils

The assembly of α-helical coiled coils from amphiphilic peptides depends subtly on the sequence of the oligomer with a variety of important biological consequences [49]. Consequently, the control over the assembly of such (designed and/or non-natural) peptides has been the subject of numerous investigations. Here we focus on the work from the laboratories of Kumar and Kennan that involves self-sorting processes. In 2001 Kumar's group reported the synthesis of two amphiphilic peptides (**H** and **F**)—one containing a hydrophobic face composed of leucine residues (**H**) and one containing hexafluoroleucine residues (**F**)—terminated in cysteine residues [50,51]. When the mixture of the fluorophobic and hydrophobic peptides were allowed to form disulfides under oxidizing conditions in water, only the **H · H** and **F · F** peptides were formed. This high fidelity self-sorting process is

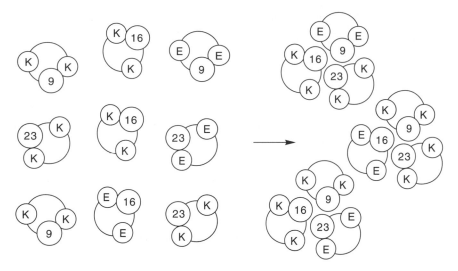

T_9E: AcNH-EMKQLEKE**X**EELESEAQQLEKEAAQLEKEVG-Am
T_9K: AcNH-KMKQLKKK**X**EELKSKAQQLKKKAAQLKKKVG-Am

$T_{16}E$: AcNH-EMKQLEKEAEELESE**X**QQLEKEAAQLEKEVG-Am
$T_{16}K$: AcNH-KMKQLKKKAEELKSK**X**QQLKKKAAQLKKKVG-Am

$T_{23}E$: AcNH-EMKQLEKEAEELESEAQQLEKE**X**AQLEKEVG-Am
$T_{23}K$: AcNH-KMKQLKKKAEELKSKAQQLKKK**X**AQLKKKVG-Am

Scheme 4.13 Self-sorting within amphiphilic peptides driven by steric matching and electrostatic effects.

driven by solvophobic effects. Schnarr and Kennan reported a very interesting system in which six different peptides spontaneously assemble into three different 1:1:1 heterotrimers in a self-sorting process (Scheme 4.13) [52]. To promote the 1:1:1 heterotrimer specificity, Schnarr and Kennan utilized a steric matching approach [53] based on the size of hydrophobic side chains (alanine versus cyclohexylalanine) at positions 9, 16, and 23 of the peptide. Only one cyclohexylalanine may reside at a given level of the interface of the 1:1:1 heterotrimer. As a secondary recognition element, the charge (e.g., lysine versus glutamic acid) at the residues adjacent (positions e and g) to the hydrophobic interface was employed. Remarkably, a 2:2:2:1:1:1 mixture of T_9K, $T_{16}K$, $T_{23}K$, T_9E, $T_{16}E$, and $T_{23}E$ forms a mixture of the three heterotrimers $T_9E \cdot T_{16}K \cdot T_{23}K$, $T_9K \cdot T_{16}K \cdot T_{23}K$, and $T_9K \cdot T_{16}E \cdot T_{23}K$ in a high fidelity self-sorting process [52]. The Kennan group has also shown how addition of specific peptides to such a mixture can be used to trigger a change in the composition [54–56] in a way similar to that reported above for the folding of non-natural oligomers inside CB[n] molecular containers.

4.6. Connections between Self-Sorting and Dynamic Combinatorial Chemistry

Dynamic Combinatorial Chemistry and self-sorting represent opposite ends of what may be considered a thermodynamic continuum. In DCC one hopes to prepare a library in which all possible members are represented and are of equal free energy. In a self-sorting system, although the number of potential species is large, only a small subset of those library members is present at significant concentrations. Both types of processes are useful, and it is perhaps unsurprising that both concepts are beginning to be employed simultaneously, as described in this section.

4.6.1. Construction, Substitution, and Sorting of Metallo-Organic Structures via Subcomponent Self-Assembly

Jonathan Nitschke's group has been particularly active recently in the development of self-sorting systems based on metal–ligand interactions. Here we describe one recent contribution to the area [57]; the reader is referred to a recent review for more comprehensive information [58]. Sarma and Nitschke report an extremely well-defined system comprising anilines **45** and **46**, aldehydes **47–49**, and Cu^+ for which they have elucidated the rules that govern the nature and concentration of all products formed (Scheme 4.14). In DMSO–CH_3CN solution, an equilibrium is established with the eight possible imines (**50–57**), indicating no innate preference exists for a specific imine. Remarkably, when Cu^+ was added to the system, a self-sorting process occurred that resulted in the formation of equimolar amounts of four discrete products (**58–61**). This system is so well defined that "more complex mixtures, having arbitrary product ratios are predicted to be readily accessible . . . by simply mixing together subcomponents in the ratio in which they are found in the desired collection of product structures" [57].

4.6.2. Self-Sorting Processes Lead to Dynamic Combinatorial Libraries with New Network Topologies

Kay Severin's group has previously investigated the preparation of dynamic combinatorial libraries of metallo-macrocycles for use as sensors for Li^+ at physiologically relevant conditions [59]. For example, they have found that ligand **62** and metal complexes **63** undergo a self-assembly process to yield the trimeric complexes depicted in Scheme 4.15 [60]. Mixtures of **64** and

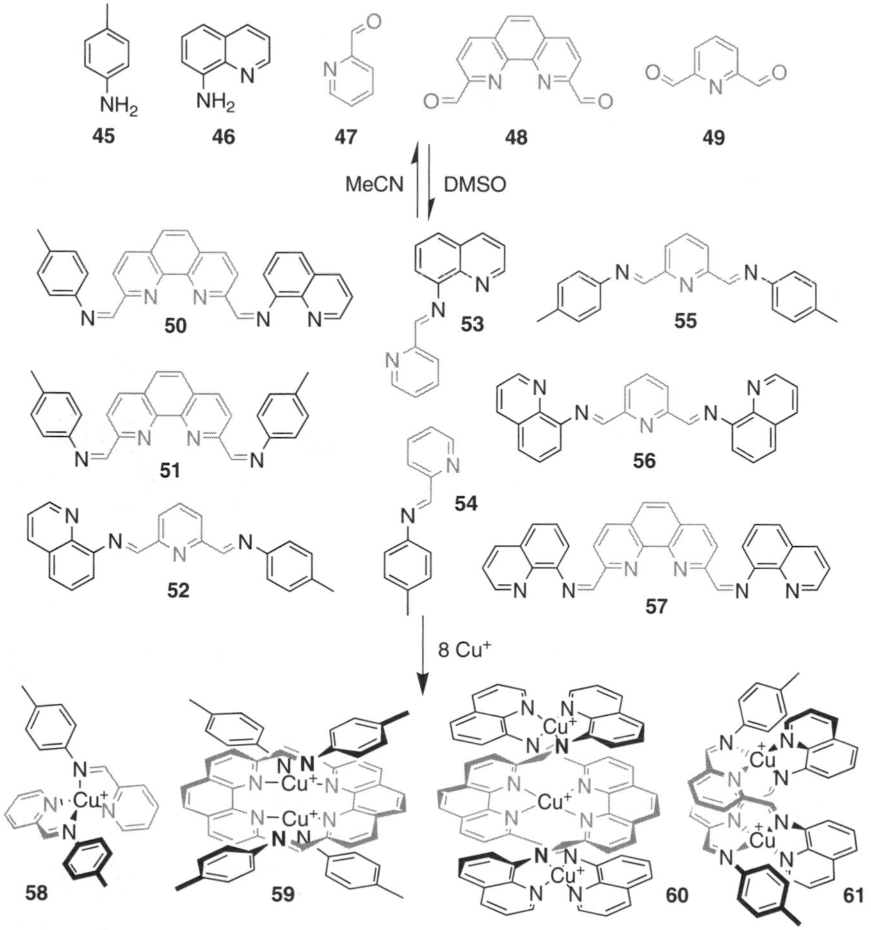

Scheme 4.14 Formation of a dynamic imine library and Cu⁺ templated self-sorting.

two different metal fragments (e.g., **63b** and **63d**) form a DCL comprising four members (**63b · 64**)$_3$, (**63b · 64**)$_2$ (**63d · 64**), (**63b · 64**) (**63d · 64**)$_2$, and (**63d · 64**)$_3$. In the presence of low concentrations of Li⁺, the dominant lithium-containing complex was (**63b · 64**)$_3$, which reflects the fact that this library member has the highest affinity for Li⁺. At higher concentrations of Li⁺, however, (**63b · 64**)$_2$ (**63d · 64**) was dominant, which reflects the fact that *more equivalents of a weaker binding complex provide more thermodynamic driving force than fewer equivalents of a tighter binding complex.* Similarly, it was possible to prepare DCLs from ligand **62** and metal fragments **63b** and **63d** that comprise all four possible metallo-macrocycles

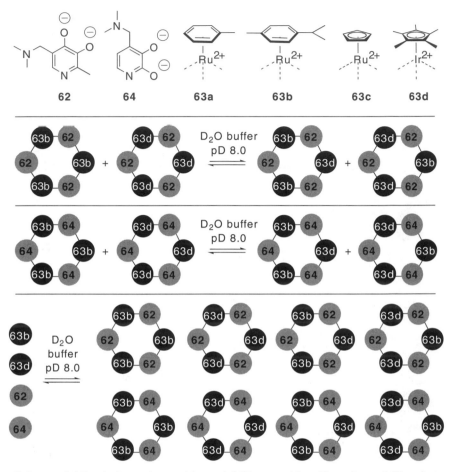

Scheme 4.15 A dynamic combinatorial library with self-sorting sublibraries.

$(63b \cdot 62)_3$, $(63b \cdot 62)_2$ $(63d \cdot 62)$, $(63b \cdot 62)$ $(63d \cdot 62)_2$, and $(63d \cdot 62)_3$. The four components of this DCL do not bind Li^+. Very interesting, however, was the behavior of a system comprising both ligands **62** and **64** and metal complexes **63b** and **63d**. Of the 24 conceivable mixed metallo-macrocycles, only eight were formed, each of which contained a single type of ligand (**62** or **64**). This self-sorting mixture consists, therefore, of the two four-component sublibraries obtained when only a single ligand is employed. It was found that the composition of the eight-component library responded differently to the templating influence of Li^+ because "the sub-library of **62** complexes can act as a reservoir for the metal fragment **63b** which is required for the formation of the high affinity receptors $(63b \cdot 64)_3$ and $(63b \cdot 64)_2$

(**63d** · **64**). The competition between (**63b** · **64**)$_3$ and (**63b** · **64**)$_2$ (**63d** · **64**) is therefore 'buffered' by the four additional DCL members [60]." The connectivity between these two sublibraries amounts to a new type of network topology that can be exploited in DCC and self-sorting systems.

4.7. Applications of Self-Sorting in Material Science and Nanotechnology

The ability of certain systems to undergo high fidelity self-sorting processes allows the precise positioning of molecules from within a complex mixture. Accordingly, there are a number of applications in materials science and nanotechnology that are enabled by self-sorting processes. This section discusses several examples of representative applications.

4.7.1. Self-Assembly of Polymers and Nanoparticles on Patterned Surfaces

Rotello and coworkers report a very interesting example of self-sorting on a patterned surface (Scheme 4.16) [61]. For this purpose, a Si wafer was spin-coated with a cationic polymer (PVMP) that was then photo-crosslinked. Patterning was performed by spin-coating a layer of Thy-PS followed by treating with UV light through a photo-resist mask. The resulting square-patterned surface contains two recognition regions: one based on PVMP electrostatic interactions and one based on the H-bonding ADA pattern of the Thy-PS polymer. Interestingly, applying solutions of either anionic nanoparticles COO-NP or diaminopyridine-derived DAD H-bonding polymer DAP-PS results in the selective derivatization of the complementary region of the surface. The simultaneous application of both DAP-PS and COO-NP results in the derivatization of both regions in a self-sorting process. The work provides a step toward the rapid, multicomponent, three-dimensional fabrication of complex functional materials.

4.7.2. Geometric Self-Sorting in DNA Self-Assembly

Deoxyribonucleic acid (DNA) is widely recognized for its ability to undergo high fidelity pairing with its cognate strands based on AT and GC base pairing preferences and its use in studies of self-assembly [62]. Chengde Mao's group recently examined whether geometric features of DNA assemblies could also be used to drive self-sorting processes (Scheme 4.17). Accordingly, they designed two related DNA tiles, namely, a four-point star **65** and

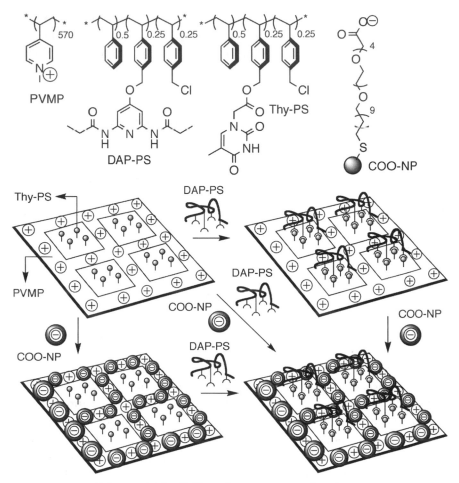

Scheme 4.16   Self-sorting on patterned surfaces.

a three-point star **66** motif. The sticky single-stranded ends of both DNA tiles are identical and self-complementary; the only difference is their relative geometrical relationship (e.g., 90° versus 120°). The three-pointed star gives rise to an extended hexagonal lattice structure **67** as determined by the atomic force microscope (AFM), whereas the four-pointed star gives rise to extended square lattice structures **68**. Interestingly, when both tiles are mixed (100:0 to 0:100 mole fraction), a single type of assembly is formed. Below a 30:70 ratio of **65** to **66** the hexagonal arrays dominate, whereas above 30:70 the square arrays are mainly observed. The work demonstrates that rational design of DNA sequences can be used to augment the range of self-sorting systems that can be prepared. The ability to

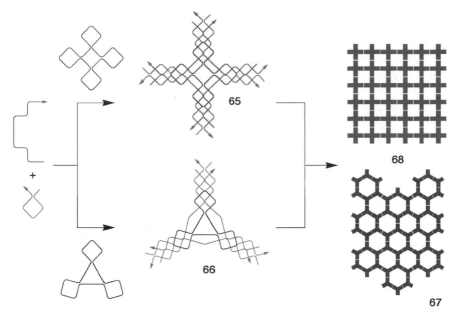

Scheme 4.17   Geometric self-sorting of DNA nanostructures.

use DNA sequences in such systems is significant because it potentially enables replication and amplification events [63–66].

4.7.3. Self-Sorting in Polymers

The group of Marcus Weck took inspiration from the self-sorting ability of one of the natural biopolymers, DNA, and decided to develop the concept of self-sorting within synthetic polymers. Their goal was the creation of a "universal polymer backbone" that could be functionalized in a subsequent orthogonal noncovalent fashion with the hope that such a system would allow the rapid generation of polymeric materials for application in materials science, drug delivery, and biomimetic chemistry. For this purpose, Burd and Weck prepared two norbornene monomers and polymerized them (random, block, homopolymers) by ring-opening metathesis polymerization to yield **69** with a range of molecular weights (12–66 kDa) and mole fractions (Scheme 4.18) [23]. A series of ^{1}H NMR titrations of these polymers with the complementary H-bonding modules **70** and **71** were performed. Remarkably, Burd and Weck found that the association constant of **70** and **71** for their H-bonding partners were comparable regardless of whether those recognition units were monomeric, homopolymeric,

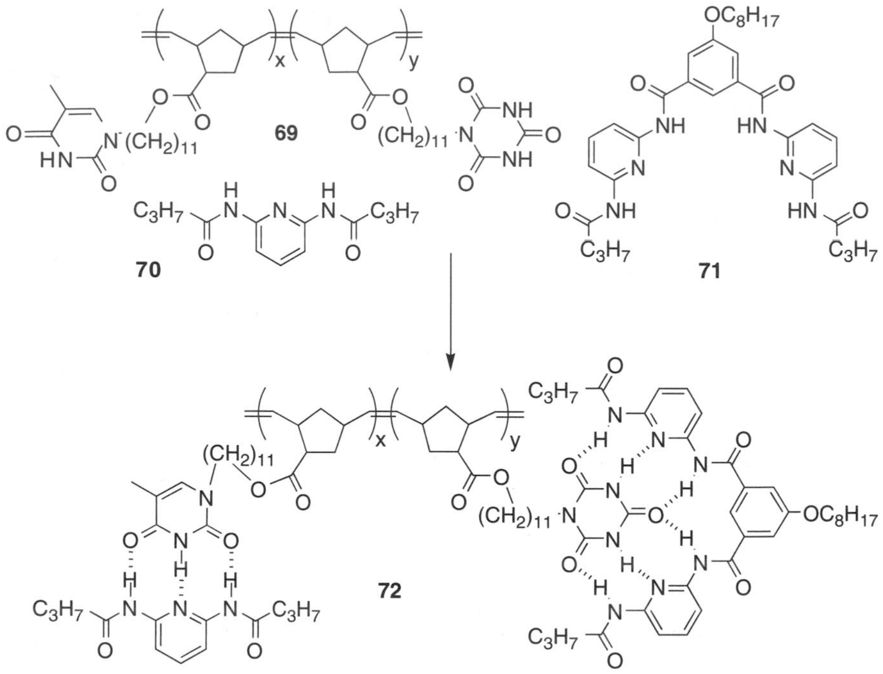

Scheme 4.18 Self-sorting noncovalent functionalization of a universal polymer backbone.

or within a block or random or block copolymer. In addition to comparable association constants, the amide region of the ^{1}H NMR spectra of **72** under saturating conditions displayed the fingerprint of self-sorting—that the NMR spectra of the mixture is equal to the sum of its parts. In related work, Weck has employed other noncovalent interactions to derivatize the side chains of related polymer backbones and plans to use them in applications such as electronic materials, polymer-mediated drug delivery, and tissue engineering.

4.7.4. Self-Sorting Gels

David Smith's group at the University of York decided to investigate the possibility of self-sorting during gelation processes. For their experiments they designed the lysine-derived first- and third-generation dendritic peptides L-**73**, D-**73**, and L-**74** with the goal of deciphering the influence of head group size and chirality on gelation (Scheme 4.19) [67]. The influence of head group size was studied with mixtures of L-**73** and L-**74**. Very

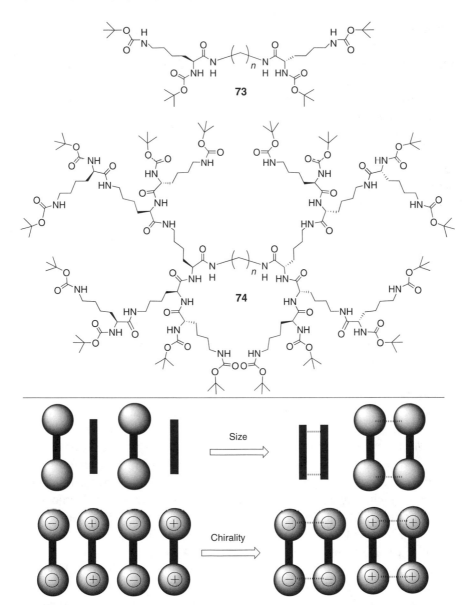

Scheme 4.19   Self-sorting in dendritic peptide gels based on head group size or chirality.

interestingly, the presence of L-**73** did not affect the gelation temperature (T_{gel}) of the more effective L-**74** gelator. Further studies with small-angle X-ray scattering and circular dichroism as the analytical tools once again indicate that L-**73** does not influence the gelation of L-**74**, which is consistent with an efficient self-sorting process governed by the size of the dendritic head group (e.g., first versus third generation). The influence of molecular chirality on gelation mixtures of L-**73** and D-**73** was studied by measuring T_{gel} as a function of mole fraction of D-**73**. T_{gel} was found to be relatively invariant as a function of mole fraction, which once again suggests a self-sorting in the gelation process. This conclusion is backed up by the results of CD measurements that show a linear dependence on mole fraction. These experiments demonstrate that subtle molecular recognition pathways can be used to control self-organization on the macroscopic scales by efficient self-sorting processes. The work is expected to enable the creation of multifunctional nanomaterials.

4.8. Conclusions

This chapter focused on the preparation of self-sorting systems. In this area of research much inspiration is drawn from living systems whose complex web of recognition events is carefully orchestrated in both time and space and do not exhibit undesired crosstalk between different pathways. The components of natural and synthetic self-sorting systems display the ability to efficiently distinguish between self and nonself and form a single set of structures in high fidelity recognition processes. As such, self-sorting processes can be viewed as residing at the opposite end of a thermodynamic continuum from idealized dynamic combinatorial libraries in which all building blocks are incorporated without selectivity. To date, some covalent-bond-forming reactions (transesterification and imine formation) and a variety of noncovalent interactions have been shown to drive self-sorting processes including H-bonds, π–π interactions, electrostatic interactions, solvophobic effects, and metal–ligand coordination. Such self-sorting processes can be divided into several categories including narcissistic versus social self-sorting and thermodynamic versus kinetic self-sorting.

A number of research groups have been involved in the development of pairs or sets of modules that undergo high fidelity self-sorting processes. Accordingly, a number of examples of both narcissistic and social self-sorting processes have been reported in which the system behaves as a linear combination of its component parts. The development of sets of such orthogonal recognition pairs are of great potential utility in a variety of application areas including supramolecular polymer chemistry, gelators, DNA nanotechnology, and surface functionalization.

Most recently, several groups have been pushing self-sorting systems into new areas by introducing new concepts derived from either biological systems or much desired nanotechnology applications. For example, the groups of Severin and Nitschke have demonstrated a link between DCC and self-sorting processes and how the behavior of such combined systems differs from their separate subsystems. Our group recently demonstrated how the high binding affinity and selectivity inherent in CB[n] receptors can be used to drive the folding, forced unfolding, and refolding of a non-natural oligomer into four distinct conformations. This stepwise process transforms one well-defined self-sorted state into another by manipulating the overall free energy of the system by the addition of new components. Related concepts have been demonstrated by Kennan in coiled-coil systems and Nitschke in the triggered interconversion of sets of metal complexes. The development of systems that can be triggered to undergo interconversion between multiple well-defined states would be very useful in the preparation and interconnection of molecular machines [13, 68, 69]. In order to endow synthetic self-sorting systems with some of the unique abilities of living systems (e.g., living far from equilibrium), we recently demonstrated an example of kinetic self-sorting. For this purpose we used guests with multiple binding epitopes. Just like living systems, which reach equilibrium only at death, the individual binding epitopes of such guests orchestrate the approach toward equilibrium exhibited by the system. Much still remains to be done. For example, the incorporation of catalytic events into self-sorting systems, and the introduction of a non-natural energy source analogous to ATP, would allow manipulation of the overall free energy of the system that could be used to drive emergent processes. Beyond that, the physical boundaries of such systems must be defined to allow restricted (privileged) interaction with the outside world and must be engineered to remain far from equilibrium. Finally, if such non-natural self-sorting systems can be interfaced with natural systems that have the ability to replicate and evolve, it might be possible to exert control over the corresponding biological system.

References

1. Cram, D. J. Molecular hosts and guests, and their complexes. *Angew. Chem. Int. Ed. Engl.* **1988**, *27*, 1009–1020.
2. Lehn, J. M. Supramolecular chemistry – Molecules, supermolecules, and molecular functional units. *Angew. Chem. Int. Ed. Engl.* **1988**, *27*, 89–112.
3. Pedersen, C. J. The discovery of crown ethers. *Angew. Chem. Int. Ed. Engl.* **1988**, *27*, 1021–1027.
4. Oshovsky, G. V.; Reinhoudt, D. N.; Verboom, W. Supramolecular chemistry in water. *Angew. Chem. Int. Ed.* **2007**, *46*, 2366–2393.

5. Schenning, A. P. H. J.; Meijer, E. W. Supramolecular electronics; nanowires from self-assembled pi-conjugated systems. *Chem. Commun.* **2005**, 3245–3258.

6. Rebek, J., Jr. Simultaneous encapsulation: Molecules held at close range. *Angew. Chem. Int. Ed.* **2005**, *44*, 2068–2078.

7. Shenhar, R.; Rotello, V. M. Nanoparticles: Scaffolds and building blocks. *Acc. Chem. Res.* **2003**, *36*, 549–561.

8. Fujita, M.; Umemoto, K.; Yoshizawa, M.; Fujita, N.; Kusukawa, T.; Biradha, K. Molecular paneling via coordination. *Chem. Commun.* **2001**, 509–518.

9. Seidel, S. R.; Stang, P. J. High-symmetry coordination cages via self-assembly. *Acc. Chem. Res.* **2002**, *35*, 972–983.

10. Wright, A. T.; Anslyn, E. V. Differential receptor arrays and assays for solution-based molecular recognition. *Chem. Soc. Rev.* **2006**, *35*, 14–28.

11. McNally, B. A.; Leevy, W. M.; Smith, B. D. Recent advances in synthetic membrane transporters. *Supramol. Chem.* **2007**, *19*, 29–37.

12. Davis, A. P.; Sheppard, D. N.; Smith, B. D. Development of synthetic membrane transporters for anions. *Chem. Soc. Rev.* **2007**, *36*, 348–357.

13. Kay, E. R.; Leigh, D. A.; Zerbetto, F. Synthetic molecular motors and mechanical machines. *Angew. Chem. Int. Ed.* **2007**, *46*, 72–191.

14. Brady, P. A.; Bonar-Law, R. P.; Rowan, S. J.; Suckling, C. J.; Sanders, J. K. M. 'Living' macrolactonization: Thermodynamically controlled cyclization and interconversion of oligocholates. *Chem. Commun.* **1996**, 319–320.

15. Huc, I.; Lehn, J.-M. Virtual combinatorial libraries: Dynamic generation of molecular and supramolecular diversity by self-assembly. *Proc. Natl. Acad. Sci. U.S.A.* **1997**, *94*, 2106–2110.

16. Klekota, B.; Hammond, M. H.; Miller, B. L. Generation of novel DNA-binding compounds by selection and amplification from self-assembled combinatorial libraries. *Tetrahedron Lett.* **1997**, *38*, 8639–8642.

17. Corbett, P. T.; Sanders, J. K. M.; Otto, S. Competition between receptors in dynamic combinatorial libraries: Amplification of the fittest? *J. Am. Chem. Soc.* **2005**, *127*, 9390–9392.

18. Corbett, P. T.; Otto, S.; Sanders, J. K. M. Correlation between host–guest binding and host amplification in simulated dynamic combinatorial libraries. *Chem. Eur. J.* **2004**, *10*, 3139-3143.

19. Severin, K. The advantage of being virtual-target-induced adaptation and selection in dynamic combinatorial libraries. *Chem. Eur. J.* **2004**, *10*, 2565–2580.

20. Saur, I.; Severin, K. Selection experiments with dynamic combinatorial libraries: The importance of the target concentration. *Chem. Commun.* **2005**, 1471–1473.

21. Kramer, R.; Lehn, J. M.; Marquis-Rigault, A. Self-recognition in helicate self-assembly: Spontaneous formation of helical metal complexes from mixtures of ligands and metal ions. *Proc. Natl. Acad. Sci. U.S.A.* **1993**, *90*, 5394–5398.

22. Rowan, S. J.; Hamilton, D. G.; Brady, P. A.; Sanders, J. K. M. Automated recognition, sorting, and covalent self-assembly by predisposed building blocks in a mixture. *J. Am. Chem. Soc.* **1997**, *119*, 2578–2579.

23. Pollino, J. M.; Stubbs, L. P.; Weck, M. One-step multifunctionalization of random copolymers via self-assembly. *J. Am. Chem. Soc.* **2004**, *126*, 563–567.

24. Lagona, J.; Mukhopadhyay, P.; Chakrabarti, S.; Isaacs, L. The cucurbit[*n*]uril family. *Angew. Chem. Int. Ed.* **2005**, *44*, 4844–4870.

25. Lee, J. W.; Samal, S.; Selvapalam, N.; Kim, H.-J.; Kim, K. Cucurbituril homologues and derivatives: New opportunities in supramolecular chemistry. *Acc. Chem. Res.* **2003**, *36*, 621–630.

26. Cai, M.; Shi, X.; Sidorov, V.; Fabris, D.; Lam, Y.-F.; Davis, J. T. Cation-directed self-assembly of lipophilic nucleosides: The cation's central role in the structure and dynamics of a hydrogen-bonded assembly. *Tetrahedron* **2002**, *58*, 661–671.

27. Shi, X.; Fettinger, J. C.; Davis, J. T. Homochiral g-quadruplexes with Ba^{2+} but not with K^+: The cation programs enantiomeric self-recognition. *J. Am. Chem. Soc.* **2001**, *123*, 6738–6739.

28. Shi, X.; Fettinger, J. C.; Cai, M.; Davis, J. T. Enantiomeric self-recognition: Cation-templated formation of homochiral isoguanosine pentamers. *Angew. Chem. Int. Ed.* **2000**, *39*, 3124–3127.

29. Wu, A.; Chakraborty, A.; Witt, D.; Lagona, J.; Damkaci, F.; Ofori, M. A.; Chiles, J. K.; Fettinger, J. C.; Isaacs, L. Methylene-bridged glycoluril dimers: Synthetic methods. *J. Org. Chem.* **2002**, *67*, 5817–5830.

30. Chakraborty, A.; Wu, A.; Witt, D.; Lagona, J.; Fettinger, J. C.; Isaacs, L. Diastereoselective formation of glycoluril dimers: Isomerization mechanism and implications for cucurbit[*n*]uril synthesis. *J. Am. Chem. Soc.* **2002**, *124*, 8297–8306.

31. Witt, D.; Lagona, J.; Damkaci, F.; Fettinger, J. C.; Isaacs, L. Diastereoselective formation of methylene-bridged glycoluril dimers. *Org. Lett.* **2000**, *2*, 755–758.

32. Rowan, A. E.; Elemans, J. A. A. W.; Nolte, R. J. M. Molecular and supramolecular objects from glycoluril. *Acc. Chem. Res.* **1999**, *32*, 995–1006.

33. (a) Isaacs, L.; Witt, D. Enantiomeric self-recognition of a facial amphiphile triggered by [{Pd(ONO$_2$)(en)}$_2$]. *Angew. Chem., Int. Ed.* **2002**, *41*, 1905–1907; (b) Wu, A.; Chakraborty, A.; Fettinger, J. C.; Flowers, R. A., II; Isaacs, L. Molecular clips that undergo heterochiral aggregation and self-sorting. *Angew. Chem., Int. Ed.* **2002**, *41*, 4028–4031; (c) Mukhopadhyay, P.; Wu, A.; Isaacs, L. Social Self-Sorting in Aqueous Solution. *J. Org. Chem.* **2004**, *69*, 6157–6164.

34. (a) Wu, A.; Isaacs, L. Self-Sorting: The Exception or the Rule? *J. Am. Chem. Soc.* **2003**, *125*, 4831–4835; (b) Liu, S.; Ruspic, C.; Mukhopadhyay, P.; Chakrabarti, S.; Zavalij, P. Y.; Isaacs, L. The Cucurbit[n]uril Family: Prime

Components for Self-Sorting Systems. *J. Am. Chem. Soc.* **2005**, *127*, 15959–15967; (c) Mukhopadhyay, P.; Zavalij, P. Y.; Isaacs, L. High Fidelity Kinetic Self-Sorting in Multi-Component Systems Based on Guests with Multiple Binding Epitopes. *J. Am. Chem. Soc.* **2006**, *128*, 14093–14102; (d) Liu, S.; Zavalij, P. Y.; Lam, Y.-F.; Isaacs, L. Refolding Foldamers: Triazene-Arylene Oligomers That Change Shape with Chemical Stimuli. *J. Am. Chem. Soc.* **2007**, *129*, 11232–11241; (e) Wu, A.; Mukhopadhyay, P.; Chakraborty, A.; Fettinger, J. C.; Isaacs, L. Molecular Clips Form Isostructural Dimeric Aggregates from Benzene to Water. *J. Am. Chem. Soc.* **2004**, *126*, 10035–10043; (f) Chakrabarti, S.; Mukhopadhyay, P.; Lin, S.; Isaacs, L. Reconfigurable Four Component Molecular Switch Based on pH-Controlled Guest Swapping. *Org. Lett.* **2007**, *9*, 2349–2352; (g) Ghosh, S.; Wu, A.; Fettinger, J. C.; Zavalij, P. Y.; Isaacs, L. Self-sorting molecular clips. *J. Org. Chem.* **2008**, *73*, 5915–5925.

35. Beijer, F. H.; Sijbesma, R. P.; Kooijman, H.; Spek, A. L.; Meijer, E. W. Strong dimerization of ureidopyrimidones via quadruple hydrogen bonding. *J. Am. Chem. Soc.* **1998**, *120*, 6761–6769.

36. Castellano, R. K.; Nuckolls, C.; Rebek, J., Jr. Transfer of chiral information through molecular assembly. *J. Am. Chem. Soc.* **1999**, *121*, 11156–1163.

37. Wyler, R.; de Mendoza, J.; Rebek, J., Jr. Formation of a cavity by dimerization of a self-complementary molecule via hydrogen bonds. *Angew. Chem. Int. Ed. Engl.* **1993**, *32*, 1699–1701.

38. Jolliffe, K. A.; Timmerman, P.; Reinhoudt, D. N. Noncovalent assembly of a fifteen-component hydrogen-bonded nanostructure. *Angew. Chem. Int. Ed.* **1999**, *38*, 933–937.

39. Taylor, P. N.; Anderson, H. L. Cooperative self-assembly of double-strand conjugated porphyrin ladders. *J. Am. Chem. Soc.* **1999**, *121*, 11538–11545.

40. Mock, W. L.; Shih, N. Y. Dynamics of molecular recognition involving cucurbituril. *J. Am. Chem. Soc.* **1989**, *111*, 2697–2699.

41. Mock, W. L.; Shih, N. Y. Structure and selectivity in host–guest complexes of cucurbituril. *J. Org. Chem.* **1986**, *51*, 4440–4446.

42. Marquez, C.; Hudgins, R. R.; Nau, W. M. Mechanism of host–guest complexation by cucurbituril. *J. Am. Chem. Soc.* **2004**, *126*, 5806–5816.

43. Marquez, C.; Nau, W. M. Two mechanisms of slow host–guest complexation between cucurbit[6]uril and cyclohexylmethylamine: pH-responsive supramolecular kinetics. *Angew. Chem. Int. Ed.* **2001**, *40*, 3155–3160.

44. Maeda, C.; Kamada, T.; Aratani, N.; Osuka, A. Chiral self-discriminative self-assembling of *meso–meso* linked diporphyrins. *Coord. Chem. Rev.* **2007**, *251*, 2743–2752.

45. Kamada, T.; Aratani, N.; Ikeda, T.; Shibata, N.; Higuchi, Y.; Wakamiya, A.; Yamaguchi, S.; Kim, K. S.; Yoon, Z. S.; Kim, D.; Osuka, A. High fidelity self-sorting assembling of *meso*-cinchomeronimide appended *meso–meso* linked Zn(II) diporphyrins. *J. Am. Chem. Soc.* **2006**, *128*, 7670–7678.

46. Hwang, I.-W.; Kamada, T.; Ahn, T. K.; Ko, D. M.; Nakamura, T.; Tsuda, A.; Osuka, A.; Kim, D. Porphyrin boxes constructed by homochiral self-sorting assembly: Optical separation, exciton coupling, and efficient excitation energy migration. *J. Am. Chem. Soc.* **2004**, *126*, 16187–16198.

47. Ma, Y.; Kolotuchin, S. V.; Zimmerman, S. C. Supramolecular polymer chemistry: Self-assembling dendrimers using the DDA AAD (GC-like) hydrogen bonding motif. *J. Am. Chem. Soc.* **2002**, *124*, 13757–13769.

48. Corbin, P. S.; Lawless, L. J.; Li, Z.; Ma, Y.; Witmer, M. J.; Zimmerman, S. C. Discrete and polymeric self-assembled dendrimers: Hydrogen bond-mediated assembly with high stability and high fidelity. *Proc. Natl. Acad. Sci. U.S.A.* **2002**, *99*, 5099–5104.

49. Lupas, A. Coiled coils: New structures and new functions. *Trends Biochem. Sci.* **1996**, *21*, 375–382.

50. Bilgicer, B.; Kumar, K. Synthesis and thermodynamic characterization of self-sorting coiled coils. *Tetrahedron* **2002**, *58*, 4105–4112.

51. Bilgicer, B.; Xing, X.; Kumar, K. Programmed self-sorting of coiled coils with leucine and hexafluoroleucine cores. *J. Am. Chem. Soc.* **2001**, *123*, 11815–11816.

52. Schnarr, N. A.; Kennan, A. J. Specific control of peptide assembly with combined hydrophilic and hydrophobic interfaces. *J. Am. Chem. Soc.* **2003**, *125*, 667–671.

53. Schnarr, N. A.; Kennan, A. J. Peptide tic-tac-toe: Heterotrimeric coiled-coil specificity from steric matching of multiple hydrophobic side chains. *J. Am. Chem. Soc.* **2002**, *124*, 9779–9783.

54. Diss, M. L.; Kennan, A. J. Orthogonal recognition in dimeric coiled coils via buried polar-group modulation. *J. Am. Chem. Soc.* **2008**, *130*, 1321–1327.

55. Schnarr, N. A.; Kennan, A. J. Sequential and specific exchange of multiple coiled-coil components. *J. Am. Chem. Soc.* **2003**, *125*, 13046–13051.

56. Schnarr, N. A.; Kennan, A. J. pH-triggered strand exchange in coiled-coil heterotrimers. *J. Am. Chem. Soc.* **2003**, *125*, 6364–6365.

57. Sarma, R. J.; Nitschke, J. R. Self-assembly in systems of subcomponents: Simple rules, subtle consequences. *Angew. Chem. Int. Ed.* **2008**, *47*, 377–380.

58. Nitschke, J. R. Construction, substitution, and sorting of metallo-organic structures via subcomponent self-assembly. *Acc. Chem. Res.* **2007**, *40*, 103–112.

59. Grote, Z.; Lehaire, M.-L.; Scopelliti, R.; Severin, K. Selective complexation of Li$^+$ in water at neutral pH using a self-assembled ionophore. *J. Am. Chem. Soc.* **2003**, *125*, 13638–13639.

60. Saur, I.; Scopelliti, R.; Severin, K. Utilization of self-sorting processes to generate dynamic combinatorial libraries with new network topologies. *Chem. Eur. J.* **2006**, *12*, 1058–1066.

61. Xu, H.; Hong, R.; Lu, T.; Uzun, O.; Rotello, V. M. Recognition-directed orthogonal self-assembly of polymers and nanoparticles on patterned surfaces. *J. Am. Chem. Soc.* **2006**, *128*, 3162–3163.

62. Seeman, N. C. An overview of structural DNA nanotechnology. *Mol. Biotechnol.* **2007**, *37*, 246–257.

63. Rozenman, M. M.; McNaughton, B. R.; Liu, D. R. Solving chemical problems through the application of evolutionary principles. *Curr. Opin. Chem. Biol.* **2007**, *11*, 259–268.

64. Kanan, M. W.; Rozenman, M. M.; Sakurai, K.; Snyder, T. M.; Liu, D. R. Reaction discovery enabled by DNA-templated synthesis and in vitro selection. *Nature* **2004**, *431*, 545–549.

65. Gartner, Z. J.; Tse, B. N.; Grubina, R.; Doyon, J. B.; Snyder, T. M.; Liu, D. R. DNA-templated organic synthesis and selection of a library of macrocycles. *Science* **2004**, *305*, 1601–1605.

66. Calderone, C. T.; Puckett, J. W.; Gartner, Z. J.; Liu, D. R. Directing otherwise incompatible reactions in a single solution by using DNA-templated organic synthesis. *Angew. Chem. Int. Ed.* **2002**, *41*, 4104–4108.

67. Hirst, A. R.; Huang, B.; Castelletto, V.; Hamley, I. W.; Smith, D. K. Self-organisation in the assembly of gels from mixtures of different dendritic peptide building blocks. *Chem. Eur. J.* **2007**, *13*, 2180–2188.

68. Kinbara, K.; Aida, T. Toward Intelligent Molecular Machines: Directed Motions of Biological and Artificial Molecules and Assemblies. *Chem. Rev.* **2005**, *105*, 1377–1400.

69. Balzani, V.; Credi, A.; Raymo, F. M.; Stoddart, J. F. Artificial molecular machines. *Angew. Chem. Int. Ed.* **2000**, *39*, 3348–3391.

Chapter 5

Chiral Selection in DCC

Jennifer J. Becker and Michel R. Gagné

5.1. Introduction

Dynamic combinatorial chemistry (DCC) is marvelously effective at discovering receptors for a broad array of analytes. The nature of the internal competition experiment ensures (normally) that the most effective binder for the analyte of interest is amplified for subsequent identification and characterization. In the context of a host–guest assembly, the issue of stereochemistry can be manifested in a number of scenarios. These include various permutations of chiral or achiral guests, along with achiral, enantiopure, or racemic dynamic library components.

For the purpose of organizing this stereoselection chapter, we will summarize the literature using the following taxonomy; first divided by the nature of the guest and second by the nature of the dynamic combinatorial library (DCL) components.

1. Achiral guest
   - Achiral DCL components
   - Racemic DCL components
   - Homochiral DCL components
2. Chiral guest
   - Achiral DCL components
   - Racemic DCL components
   - Homochiral DCL components

Dynamic Combinatorial Chemistry, edited by Benjamin L. Miller
Copyright © 2010 John Wiley & Sons, Inc.

From the viewpoint of stereochemistry, Sections 5.2–5.4 focus on the templated/amplified components of the DCL itself, with the emphasis on how diastereomeric receptors differentially respond to the host–guest assay. Spectacular structures have emerged from the internal competition for the best binder(s) to an analyte.

When the analyte is chiral, however, the stereochemical issues evolve to how a chiral host interacts with a chiral guest. The nature and magnitude of these diastereomeric interactions ultimately control the DCL evolution and what types of hosts are amplified.

This chapter focuses on the literature from 1996 to 2007, a period of time wherein the concepts of DCC were consciously examined as a tool for chemical discovery. The aim of the authors was not to exhaustively document each instance where a chiral selection led to the diastereoselective templating of a complex structure.

5.2. Achiral Guest: Achiral DCL Component

One of the first instances of the chiroselective amplification of an achiral library component was observed in the work of Lehn and coworkers on the anion-dependent aggregation state of tris-bipy Fe(II) circular helicate complexes $[n]\mathbf{cH}$ [1]. Combinations of the linear (achiral) tris-bipyridyl ligand with Fe(II) sources provided cyclic products whose nuclearity depended on the counterion. When $FeCl_2$ was utilized, a pentanuclear structure emerged as the sole product ($[5]\mathbf{cH}$), with the "doughnut hole" of the ring being filled by a tightly bound Cl^- (Fig. 5.1). When larger anions such as BF_4^-, SO_4^{2-}, or SiF_6^{2-} were utilized, a hexanuclear structure ($[6]\mathbf{cH}$) with a concomitantly larger hole was obtained, and with Br^- a mixture of cyclic pentamer and hexamer resulted. These species were characterized by mass spectrometry (MS) and NMR, with the latter revealing a single type of Fe-center and a symmetric arrangement of ligands. NOE measurements indicated that each metal center was ligated to three bipy rings, consistent with a scenario wherein two terminals and one central bipy from three separate ligands were arranged about each iron center. The high molecular symmetry additionally required that each metal center have the same chirality (Λ or Δ), that is, be homochiral and have the same sense of helicity for each metal. Indirect evidence for this scenario was obtained from a more flexible tris-bipy ligand with a CH_2OCH_2 linkage separating the central bipy from the terminal bipys. This ligand generated a similarly symmetrical cyclic tetramer ($[4]\mathbf{cH}$), which was structurally characterized by X-ray methods.

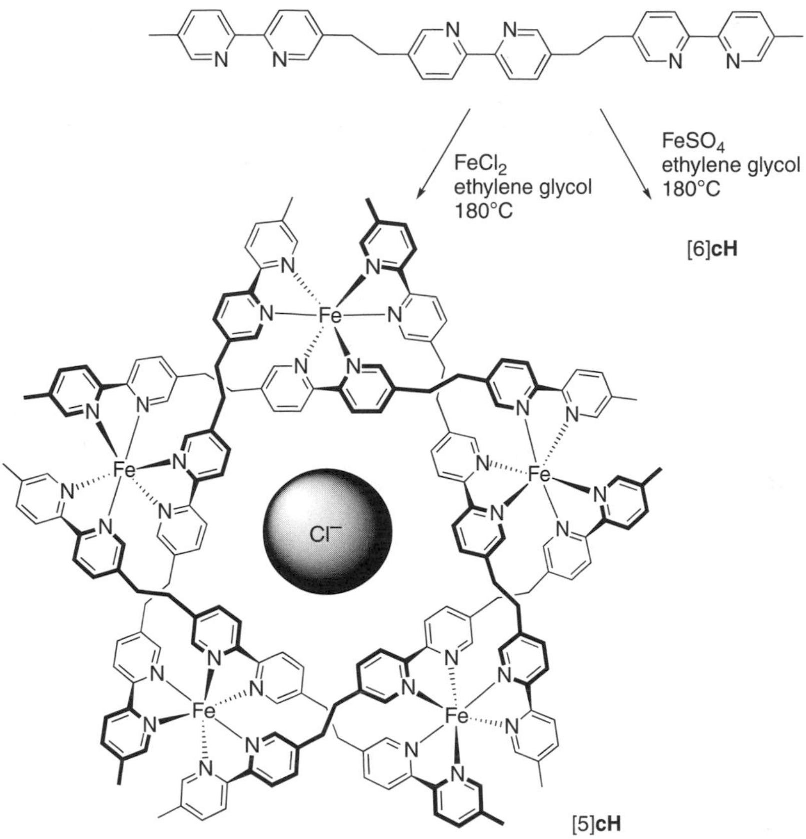

Figure 5.1   Diastereoselective assembly of chiral polynuclear Fe(II)-bipyridyl structures.

5.3. Achiral Guest: Racemic DCL Components

A recent example of diasteromeric amplification with achiral guests and a racemic library can be seen in the work of Iwasawa and coworkers. The library members consisted of a racemic polyol and 1,4-benzenedi(boronic acid) [2]. When these components were mixed in an equimolar ratio in methanol, a precipitate formed, which was insoluble in other organic solvents and thought to be a polymeric boronate. However, when the library members were mixed in the presence of toluene or benzene, a precipitate again formed, but it was soluble in several (nonprotic) organic solvents where boronic ester exchange is slow. With toluene a [2:2] complex of the polyol and diboronic acid formed, as evidenced by NMR and FAB-MS data. X-ray crystallography confirmed that a cyclic structure formed with one toluene guest molecule in the interior. Each of the templated [2:2] complexes was homochiral in nature, with the single crystal also containing the enantiomeric [2:2] species (Fig. 5.2). When benzene was used as the guest, a [3:3] complex was preferentially formed. Analysis by ¹H and ¹³C NMR indicated that this [3:3] host was constructed with three tetrols, two of one enantiomeric tetrol and one of its antipode. Other polycyclic aromatic compounds were capable of similarly templating the assembly of discrete chiral structures.

Otto, Sanders, and coworkers have utilized disulfide exchange to generate dynamic libraries of diastereomeric receptors [3]. DCLs made from a racemic dithiol led to numerous cyclic structures including four cyclic tetramers, with the *RR,RR,RR,RR* diastereomer being the most stable (along with its all-*S* enantiomer). Upon addition of N(CH₃)₄I, the meso-diastereomer shown below was amplified 400-fold (Fig. 5.3). The structure of the diastereomer was confirmed by NMR and re-equilibration

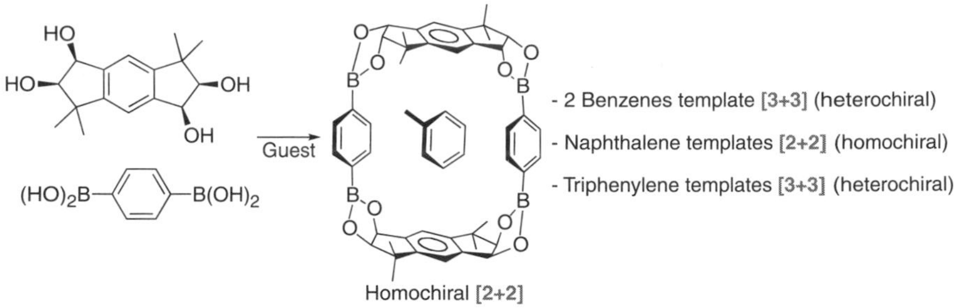

Figure 5.2 Self-assembly of chiral tetra boronic ester clathrates from a racemic tetrad.

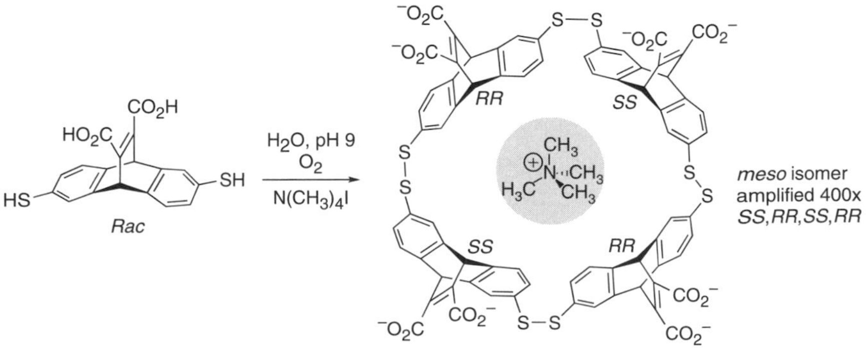

Figure 5.3 Diastereoselective amplization of an alternating, meso arrangement of dithiols to create a high affinity receptor for N(CH₃)₄⁺.

experiments. Of the four possible tetramer diastereomers, the amplified compound was highly symmetric and showed four different C–H environments in the ^{1}H NMR, thereby eliminating diastereomers *RR,RR,SS,SS* and *RR,RR,RR,SS* based on symmetry. To distinguish between the remaining two diastereomers *RR,RR,RR,RR* and *RR,SS,RR,SS*, the amplified diastereomer was isolated and dissolved under equilibration conditions. After several days the cyclic trimer *RR,RR,SS* was observed as a single diastereomer. It was presumed that in the early stages of re-equilibration, the easiest path for conversion between the tetramer and trimer is through loss of a single unit, leading to the conclusion that the observed trimer had been formed from the alternating achiral tetramer. Isothermal titration microcalorimetry indicated that the host–guest binding was strongly driven by enthalpy and cation–π interactions.

Alfonso, Gotor, and coworkers have also affected diasteroeselective amplification from racemic libraries and achiral guests [4]. Mixtures of cyclohexyl-1,2-diamine (*rac*) and 2,6-diformyl-pyridine led to a mixture of homo- and heterochiral dimers, trimers, and tetramers (Fig. 5.4). The addition of Ba(II) slightly amplified the homochiral over the heterochiral dimer (1.6:1). Templating with Cd(II) instead led to a preponderance of trimers, as evidence by ESI-MS. However, the large number of signals in the ^{1}H and ^{13}C NMR spectra indicated a heterochiral trimer. Amplification of the C_2-symmetric heterochiral trimer (Fig. 5.4) was confirmed by

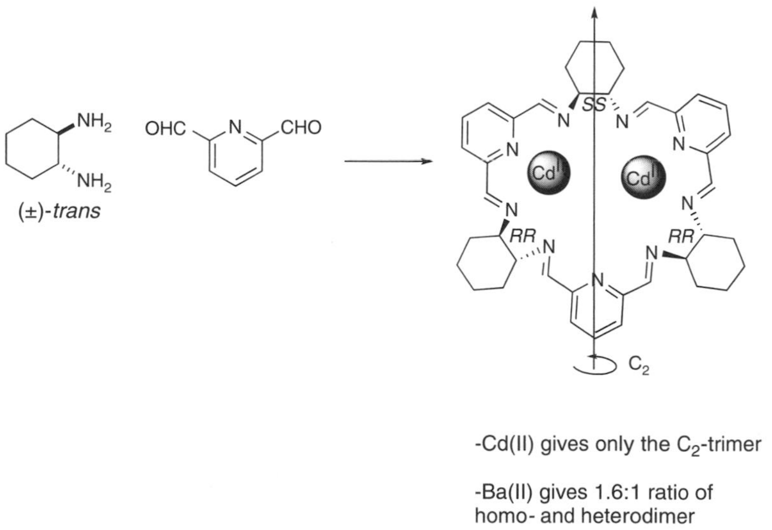

-Cd(II) gives only the C_2-trimer

-Ba(II) gives 1.6:1 ratio of homo- and heterodimer

Figure 5.4 Cd(II) templated synthesis of a C_2-symmetric receptor from a racemic diamine.

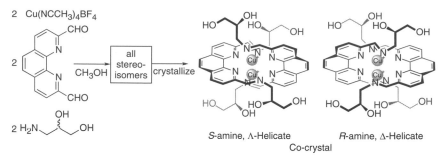

S-amine, Λ-Helicate   R-amine, Δ-Helicate
Co-crystal

Figure 5.5 Precipitation driven self sorting into matched central and axial chiralities.

comparison to independently synthesized homochiral structures. Molecular modeling indicated that the amplified species was likely a dinuclear Cd(II) metal complex.

The notions of DCC have also been used in examining the stereoselective crystallization of racemic mixtures. Nitschke has reported DCLs made with a phenanthroline, a chiral amine, and $Cu(NCCH_3)_4BF_4$ as shown in Fig. 5.5 [5]. When enantiopure amine was used, a single product was obtained. NMR and circular dichroism confirmed a diastereomerically pure dinuclear complex that matched the central and axial chiralities. When the racemic amine was used, however, a complex mixture of diastereomers was observed in solution. Upon standing, X-ray quality crystals formed, which revealed that only one diastereomer and its enantiomer were combined into distinct columns of the Δ and Λ enantiomers. The homochiral molecules were linked together through H-bonding of the hydroxyl groups. Redissolution in water gave a ^1H NMR spectrum that was the same as the original mixture of diastereomers, indicating that the resolution was driven by the crystallization, and that it was not temporally stable.

DCLs have also been made with two different types of reversible bonds, disulfides, and hydrogen bonds [6]. In this example from Sijbesma and Meijer, dimerization of the racemic thiol created a disulfide with functional groups capable of intermolecular H-bonding through a donor–donor–acceptor–acceptor (DDAA) H-bonding array (Fig. 5.6). Above 120 mM linear oligomers dominate, but below this critical concentration the speciation shifted to cyclic dimers. The ^1H NMR of a mixture of *RR* and *SS* disulfide was identical to that of an enantiopure disulfide, indicating that the *RR* disulfide only H-bonded to another *RR* disulfide, that is, a homochiral dimer. When the disulfide mixture contained both *rac-* and *meso-* diastereomers, NMR analysis showed that the *RR* and *SS* again paired in a homochiral sense, while the *RS* disulfide also paired with itself; no

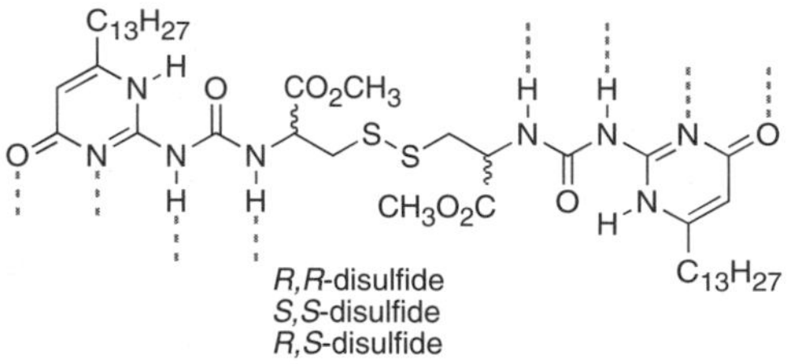

Figure 5.6 Non-covalent interactions drive a diastereoselective assembly under disulfide exchange conditions.

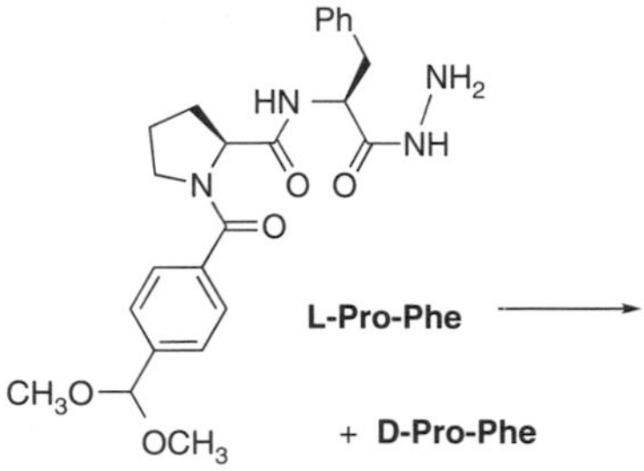

Figure 5.7 Oligomer dependent self sorting of diastereomeric building blocks.

evidence for cross pairing was observed. Under disulfide exchange conditions the *rac–meso* ratio of the disulfide changed from an initial point of 46:54 to 24:76, suggesting that the paired heterochiral structure was slightly favored over the homochiral arrangements.

Sanders and coworkers have examined the DCLs that resulted from mixing L- and D-amino acid building blocks [7]. Equilibration of the racemic mixtures led to homo- and heterochiral dimers, trimers, and tetramers depending on the amino acid building block (Fig. 5.7). When the D- and L-amino acids were of the same structure or from the same "family" (similarity of the amino acid side chain; 1°, 2°, or 3° β-carbon), the mixed trimer was more abundant than the homotrimer. When the D- and L-amino acids were from different families, self-sorting became more prevalent and the abundance of the mixed oligomers was lower. Interestingly, pairs of DCL components could be made to react like pseudo-enantiomers when the enantiomeric partner was in the same family, that is, D-monomer is in the same family as the L-monomer. These studies demonstrated the degree of library diversity that could be achieved with stereochemical mixtures of monomers.

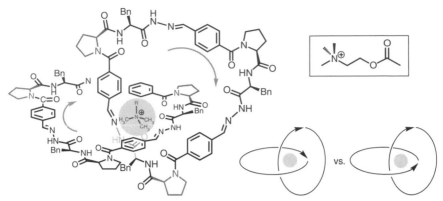

Figure 5.8 A single enchainment stereo isomer is obtained in the acetylcholine templated synthesis of a [2] catenane.

5.4. Achiral Guest: Homochiral DCL Components

In one of the most spectacular cases of a template-induced asymmetric synthesis, Sanders and coworkers combined an achiral guest and a homochiral library component to yield a unique 3 · 3 catenane [8]. Addition of acetylcholine to the normal library of cyclic oligomers caused the slow growth of a new product that was isomeric to the hexamer. The new species and the cyclic hexamer displayed the same molecular ion, but their MS/MS fragmentation patterns were distinguishing (Fig. 5.8). The cyclic hexamer sequentially fragments to the $(n - 1)$-*mer* (5-*mer*, 4-*mer*, 3-*mer*, etc.) in contrast to the new species which fragmented directly to the trimer, a phenomenon typical of [2]-catenanes with interlocked cyclic trimers (confirmed by NMR). Two possible diastereomeric catenanes were possible since each cyclic trimer is chiral; however, the NMR pointed to a single diastereomer being amplified.

Another example of an achiral guest and a homochiral library was described in the work of Williams [9], which combined Co(II), bipy, and the homochiral ligand bipy*. All possible combinations of Δ- and Λ-[Co(bipy*)$_x$(bipy)$_{3-x}$]$^{2+}$ complexes were formed as confirmed by NMR and ES-MS. Treating with D-Cl was found to shift the speciation to only [Co(bipy)$_3$]$^{2+}$ and Δ-[Co(bipy*)$_3$]$^{2+}$ (by NMR). Previous work on similar systems had shown that D-Cl deuterated the basic amine sites, and the resulting ammonium ions cooperatively bound two chloride ions [10]. In this example, addition of the achiral chloride ion amplified a single diastereomer of Δ-[Co(bipy*)$_3$]$^{2+}$ (Fig. 5.9).

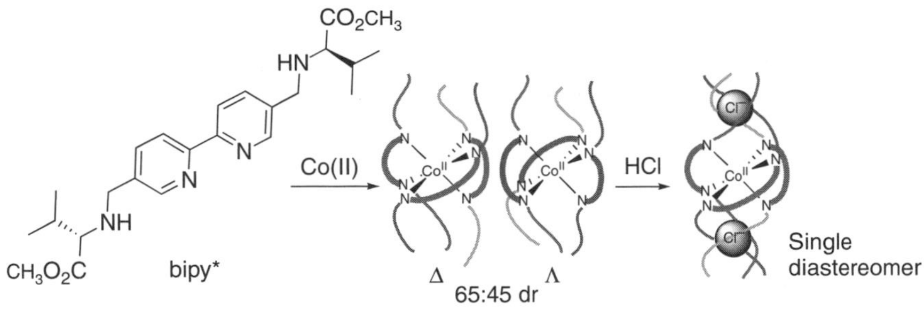

Figure 5.9   Chloride templated inter-ligand exchange to converge to a single diasteromer of product.

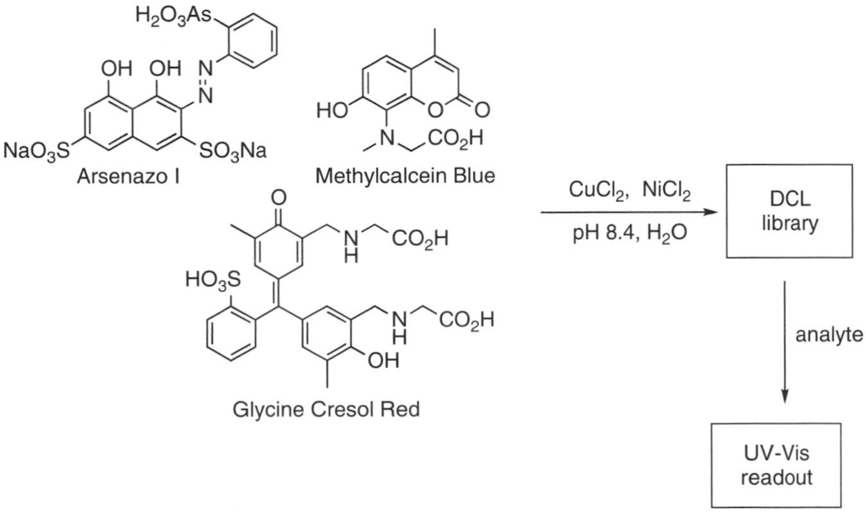

Figure. 5.10   Dynamic assembly of metal and dye (ligands) capable of diztinguishing, peptides via UV-Vis response.

5.5. Chiral Guest: Achiral DCL Component

Severin and Buryak have utilized mixtures of commercial dyes and simple metals [Cu(II) and Ni(II)] to generate DCLs capable of distinguishing closely related di- and tripeptides [11]. The DCL was generated by the combination of metal and dye, which created a complex mixture of uniquely UV-Vis-absorbing coordination compounds. Addition of the analyte (usually a di- or tripeptide) shifted the library speciation that then created a new UV-Vis spectrum (Fig. 5.10). Combining this principle with linear discriminant analysis of the spectrum (in conjunction with "learning" datasets) enabled various target peptides to be unambiguously identified. The discriminating power of this method was remarkable as regio isomers such as Ala-Phe/Phe-Ala and His-Gly-Gly/Gly-His-Gly/Gly-Gly-His were readily distinguished. Similarly impressive was the ability to recognize diastereomeric analytes, for example, L-Phe-Ala and D-Phe-Ala.

5.6. Chiral Guest: Racemic DCL Component

Gagné and coworkers utilized this combination to discover enantioselective receptors for (−)-adenosine [12]. A racemic dipeptide hydrazone [(±)-pro-aib] generated a stereochemically diverse DCL of *n-mer*. The dimers were composed of two chiral (*DD/LL*) and one achiral isomer (*DL*), the four trimers (*DDD, LLL, DDL*, and *LLD*), the tetramers of four chiral and two achiral isomers, etc. Two techniques were used to measure the enantio-imbalance that was caused by the enantioselective binding of the chiral analyte to the enantiomeric receptors (Fig. 5.11). Since the unperturbed library is optically inactive, the optical enrichment of each library component could be measured by a combined HPLC optical rotation detection scheme (laser polarimeter, LP). LP detection differentiated unselective binding (amplification but not optical enrichment) from enantioselective recognition of the analyte (amplification and optical enrichment). In this manner the *LL* dimer (*SS*) of the dipeptide was amplified and identified as the enantioselective match for (−)-adenosine.

An isotopic labeling scheme based on pseudo-enantiomers that enabled the diastereomeric receptors to be individually addressed (by LC-MS) was also examined. This methodology enabled the direct identification of the amplified diastereomer and the measurement of its selectivity over the competing stereoisomers (Fig. 5.12).

5.7. Chiral Guest: Homochiral DCL Component

The amplification of two different macrocyclic receptors for similar diastereomeric compounds has been observed by Sanders and coworkers [13]. A dynamic system of macrocyclic polyhydrazones ranging from dimer to at least undecamer was prepared from the indicated homochiral hydrazone. Templating with quinine caused the percentage of cyclic tetramer to increase from 63% to 91%. The same experiment with quinidine, a diastereomer of quinine, instead amplified the dimer from 9% to 45% (Fig. 5.13). These species were characterized by HPLC and ESI MS. Association

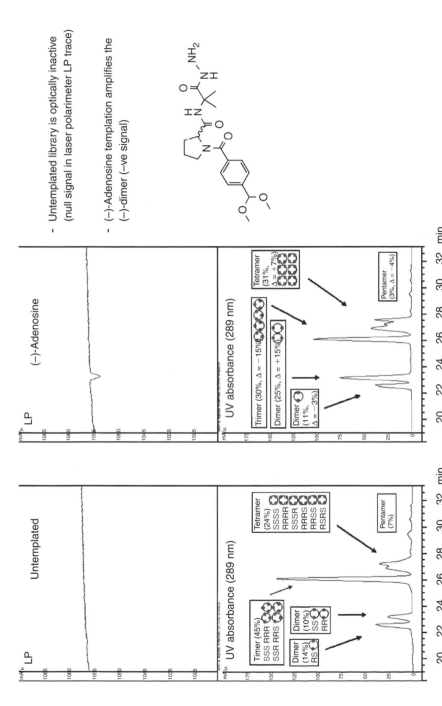

Figure 5.11 Laser polarimeter, (LP) detection to identify cases of enantioselective amplification from optically inactive DCLS.

165

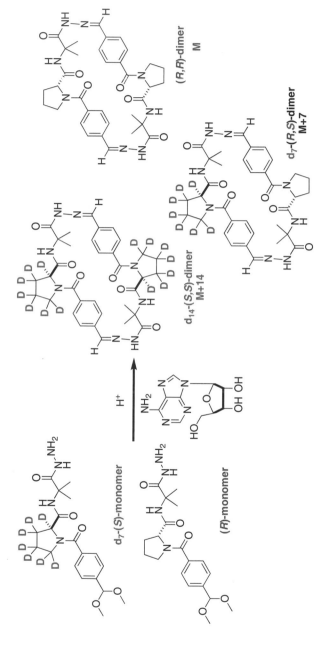

Figure 5.12 A pseudo-enantiomeriz scheme for utilizing mass spec to identify and quantify enantiomeric imbalances in a DCL.

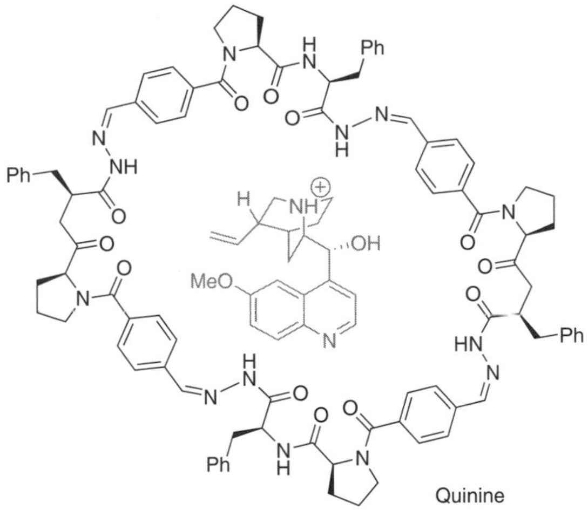

Figure 5.13   Amplification of a tetrameric receptor with quinine. The isomeric guinidine amplifies a dimer.

constants were estimated to be ~10^4 M^{-1} since the template could alter the population distribution of the library in millimolar solutions.

5.8. Summary

The rapid adoption of DCC by the chemical community bodes well for continued development and enhancement over the coming years. It stands to reason that the importance of stereochemical selectivity in a broad array of research problems (catalysis, sensors, etc.) will drive the evolution of the topic of this chapter to higher and higher levels of sophistication.

Acknowledgment

The US Army Research Office is gratefully acknowledged for financial support of the MRG contributions to this chapter (W911NF-06-1-0169).

References

1. Hasenkopf, B.; Lehn, J.-M.; Boumediene, N.; Dupont-Gervais, A.; Dorsselaer, A. V.; Kneisel, B.; Fenske, D. Self-assembly of tetra- and hexanuclear circular helicates. *J. Am. Chem. Soc.* **1997**, *119*, 10956–10962.

2. Iwasawa, N.; Takahagi, H. Boronic esters as a system for crystallization-induced dynamic self-assembly equipped with an "on–off" switch for equilibration. *J. Am. Chem. Soc.* **2007**, *129*, 7754–7755.

3. Corbett, P. T.; Tong, L. H.; Sanders, J. K. M.; Otto, S. Diastereoselective amplification of an induced-fit receptor from a dynamic combinatorial library. *J. Am. Chem. Soc.* **2005**, *127*, 8902–8903.

4. González-Álverez, A.; Alfonso, I.; Gotor, V. Highly diastereoselective amplification from a dynamic combinatorial library of macrocyclic oligoimines. *Chem. Commun.* **2006**, 2224–2226.

5. Hutin, M.; Cramer, C. J.; Gagliardi, L.; Shahi, A. R. M.; Bernardinelli, G.; Cerny, R.; Nitschke, J. R. Self-sorting chiral subcomponent rearrangment during crystallization. *J. Am. Chem. Soc.* **2007**, *129*, 8774–8780.

6. (a) ten Cate, A. T.; Dankers, P. Y. W.; Sijbesma, R. P.; Meijer, E. W. Disulfide exchange in hydrogen-bonded cyclic assemblies: Stereochemical self-selection by double dynamic chemistry. *J. Org. Chem.* **2005**, *70*, 5799–5803. (b) ten Cate, A. T.; Dankers, P. Y. W.; Kooijman, H.; Spek, A. L.; Sijbesma, R. P.; Meijer, E. W. Enantioselective cyclization of racemic supramolecular polymers. *J. Am. Chem. Soc.* **2003**, *125*, 6860–6861.

7. Liu, J.; West, K. R.; Bondy, C. R.; Sanders, J. K. M. Dynamic combinatorial libraries of hydrazone-linked pseudo-peptides: Dependence of diversity on building block structure and chirality. *Org. Biomol. Chem.* **2007**, *5*, 778–786.

8. Lam, R. T. S.; Belenguer, A.; Roberts, S. L.; Naumann, C.; Jarrosson, T.; Otto, S.; Sanders, J. K. M. Amplification of acetylcholine-binding catenanes from dynamic combinatorial libraries. *Science* **2005**, *308*, 667–669.

9. Telfer, S. G.; Yang, X.-J.; Williams, A. F. Complexes of 5,5'-aminoacido-substituted 2,2'-bipyridyl ligands: Control of diastereoselectivity with a pH switch and a chloride-responsive combinatorial library. *J. Chem. Soc. Dalton Trans.* **2004**, 699–705.

10. Telfer, S. G.; Bernardinelli, G.; Williams, A. F. Iron and cobalt complexes of 5,5'-di(methylene-*N*-aminoacidyl)-2,2'-bipyridyl ligands: Ligand design for diastereoselectivity and anion binding. *J. Chem. Soc. Dalton Trans.* **2003**, 435–440.

11. (a) Buryak, A.; Severin, K. Dynamic combinatorial libraries of dye complexes as sensors. *Angew. Chem. Int. Ed.* **2005**, *44*, 7935–7938. (b) Buryak, A.; Severin, K. Easy to optimize: Dynamic combinatorial libraries of metal–dye complexes as flexible sensors for tripeptides. *J. Comb. Chem.* **2006**, *8*, 540–543.

12. Voshell, S. M.; Gagné, M. R. The discovery of an enantioselective receptor for (−)-adenosine from a racemic dynamic combinatorial library. *J. Am. Chem. Soc.* **2006**, *128*, 12422–12423.

13. Bulos, F.; Roberts, S. L.; Furlan, R. L. E.; Sanders, J. K. M. Molecular amplification of two different receptors using diastereomeric templates. *Chem. Commun.* **2007**, 3092–3093.

Chapter 6

Dynamic Combinatorial Resolution

Marcus Angelin, Rikard Larsson, Pornrapee Vongvilai, Morakot Sakulsombat, and Olof Ramström

6.1. Introduction

Since its formulation in the mid-1990s, dynamic combinatorial chemistry (DCC) has successfully expanded to strongly influence a large variety of research areas. This success can partly be explained by the relative simplicity of the concept. In DCC, the basic tool set consists of building blocks with the ability to participate in an elementary *reversible* organic transformation. These simple tools then work together, empowered by the laws of thermodynamics, to form a dynamic library of continually interchanging species. This library is then subjected to thermodynamic or kinetic pressure through some type of screening/selection event. The dynamic library responds, eventually selecting the best components for that particular application.

6.1.1. Approaches to DCC

Looking at DCC from a general point of view, it can be divided into a limited number of approaches, all of which having their advantages and drawbacks and addressing specific challenges. They share a common feature in the initial reversible library generation step, but differ in the screening/selection phase. These approaches are described thoroughly in other chapters, but will be mentioned here briefly for clarifying reasons.

Dynamic Combinatorial Chemistry, edited by Benjamin L. Miller
Copyright © 2010 John Wiley & Sons, Inc.

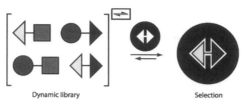

Dynamic library   Selection

Figure 6.1 The stoichiometric DCC approach.

The Stoichiometric Approach (Fig. 6.1)

Stoichiometric dynamic combinatorial libraries (DCLs) were the original DCC concept, where the dynamic generation of library constituents was performed in presence of the selection target, in the same compartment. Here, all dynamic features of the system are utilized and the dynamics allows for adaptation to internal changes or external stimuli. The adaptability, enabled by the reversibility of the processes, also makes amplification through Le Châtelier's principle possible. If one component interacts better with the target, it will be 'removed' from the equilibrating pool. This causes the system to respond by producing more of that component, a feature referred to as amplification.

The Pre-Equilibrated Approach (Fig. 6.2)

When using pre-equilibrated DCLs, library generation and screening are treated as two separated processes. The dynamic library is established under reversible conditions and followed by a subsequent screening step under static conditions. This approach does not require stoichiometric amounts of the target, which makes it suitable when working with sensitive biological targets of low abundance. A second reason to use this approach is if the connecting reaction requires conditions that are incompatible with the target. However, the static selection conditions prevent the possibility of amplification effects. This makes analysis more difficult, sometimes requiring dynamic deconvolution strategies. In this case, smaller sublibraries are constructed, screened, and compared to eventually identify the selected components.

The Iterative Approach (Fig. 6.3)

Pre-equilibrated protocols can enable constituent amplification when run iteratively. This is referred to as the iterative approach. The DCL is generated under dynamic conditions, and then subjected to static selection. After static selection, however, the selected components must be removed, for example, by using immobilized targets. The unbound species are then

Complete system

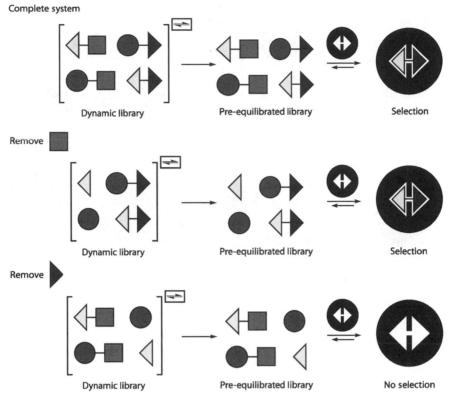

Figure 6.2   The pre-equilibrated approach and consecutive dynamic deconvolution.

retransferred to the original chamber, rescrambled, and again subjected to selection. After several rounds, one or several accumulated species may be easily identified.

The Pseudodynamic (Deletion) Approach

A related approach is based on the deletion of unbound constituents from a combinatorial library. Binding constituents are partially shielded from this event, causing the ratio of good to poor binders to increase. In this approach, library formation and deletion are two separate irreversible events, rendering a process that is formally nondynamic.

6.1.2. Applications of DCC

Due to the development of DCC and its cleverly designed approaches, scientists have been able to employ it for a large variety of applications within many separate research areas.

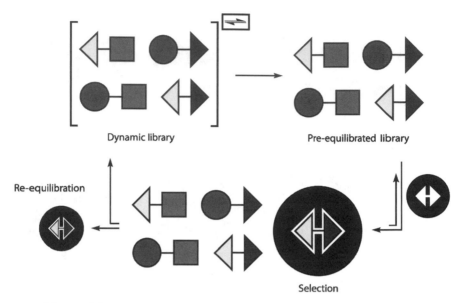

Figure 6.3   Pre-equilibrated DCL generation and iterative screening.

Synthetic Receptors and Catalysts

The development of synthetic receptors is an area known for involving lengthy, iterative processes of design, synthesis and evaluation, which has proven in many cases to be ideal for employing DCC. Due to frequently dealing with molecules of large molecular weights and complexity, the design and synthesis of entire receptors are, in the normal case, very encumbering. Instead, DCC only requires synthesis of individual fragments, which then are assembled reversibly in the presence of target molecules. In this way, the target molecule itself is allowed to thermodynamically design the ideal size and composition of the receptor. The same strategy can also be applied to construct new catalysts through dynamic affinity studies of transition-state analogs (TSAs).

DCC in Folding and Aggregation Processes

DCC can also be used to investigate secondary and tertiary folding patterns in synthetic polymers, oligomers, proteins, and nucleic acids. In this case, there is no external template molecule that governs the process, but rather a type of intramolecular recognition. However, another related process is aggregation, where self-assembly of some of the DCL components into aggregates influence library composition and amplifies members that form the thermodynamically most favored aggregate.

Ligands for Biomolecules

One of the applications where DCC has the largest potential is the discovery of new ligands for biomolecules in general and drug discovery in particular. Biomolecules are often both the most interesting and the most challenging targets. There is a large interest in finding new and more efficient ways to more quickly and effectively find specific inhibitors, substrates, or affinity analogs. Unfortunately, the biomolecules are in many cases available only in limited amounts and require careful handling and mild conditions. Their function is also often limited to aqueous environments, operating in carefully buffered solutions. This is not compatible with traditional organic chemistry, which often employs strong acids, bases, and an organic solvent as reaction medium. However, even though biomolecules greatly restrict the tools by which the dynamic combinatorial system can be constructed, there are still several examples of reversible reactions and systems that have been successfully applied to biomolecules such as enzymes, lectins, and oligonucleotides, among others.

6.2. Dynamic Combinatorial Resolution (DCR)

Dynamic combinatorial chemistry protocols are generally separated into two conceptual stages: library generation, where the selected building blocks interact reversibly (under thermodynamic control) to form a DCL, and library screening/selection, where the library is subjected to a "selection pressure," often through interactions with a provided target molecule. This is undoubtedly the major advantage of dynamic systems, in principle resulting in adaptivity to the pressure exerted.

In the vast majority of literature reporting DCC systems, the subsequent selection pressure is of thermodynamic character. In a typical case, a dynamic library becomes distorted through favorable noncovalent interactions of one or more of the library constituents with a target, which could be anything from a metal ion to a receptor protein. The distortion could also be due to internal stabilization such as specific folding or aggregation processes. These favorable interactions stabilize the particular constituent, bringing it to a lower energy state. The library then, due to its purely dynamic character, recognizes a new global energy minimum and adjusts its composition to reach it. This leads to increased formation of the stabilized component or, in other words, amplification (Fig. 6.4).

There are, however, situations where classic strategies of thermodynamically controlled screening are difficult to apply. Many times, the complex libraries are difficult to analyze in spite of the progress in analysis techniques and instrumentation. Isolation of the amplified products is also

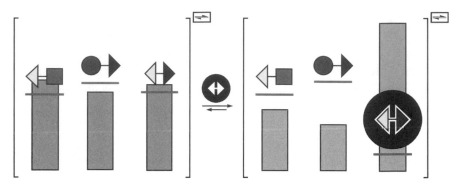

Figure 6.4   Amplification of a library member due to its favorable interaction with the target molecule.

problematic and generally requires freezing of the system dynamics, something that cannot easily be achieved in the normal case. Also, the amplified library members are, in many cases, not individually stable without the presence of the target, thereby making them even harder to isolate and characterize. Isolation of the components in an intact complex with the target is a possibility, but often difficult and generally requires the target to be of heterogenic character through, for example, immobilization on solid support, a strategy that has proved to be very difficult. Sometimes even more problematic is the fact that thermodynamic screening of dynamic libraries generally requires stoichiometric amounts of the target molecule to obtain notable changes in library composition. In the case of enzymes or receptors, and for possible applications in drug discovery, this may be impossible to obtain. If, on the other hand, the binding event is coupled to a kinetically controlled secondary reaction, only a catalytic amount of target species is required. This also may result in an individually stable and distinct compound, easy to isolate and characterize. Recently, the concept of DCR was introduced for kinetically resolved libraries (Fig. 6.5). Here, the thermodynamically controlled dynamic library is resolved by an irreversible, and selective, secondary reaction. This principle has already been demonstrated in the screening of enzyme substrates, using enzymes to select and act upon the best substrates in DCLs [1–4]. A closely related concept, termed internal dynamic combinatorial resolution (iDCR) has also recently been introduced and demonstrated (Fig. 6.6) [5,6]. In this case, an internal selection pressure from a subsequent intramolecular tandem reaction influenced the thermodynamically controlled library and caused a complete amplification effect. Broadening of this concept could have intriguing potential applications such as providing a route to systematization of reaction discovery, where novel reactions could be identified from isolation of unexpectedly

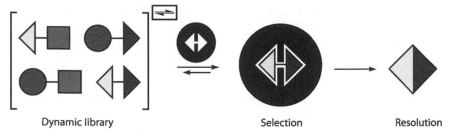

Dynamic library   Selection   Resolution

Figure 6.5 Dynamic combinatorial resolution (DCR).

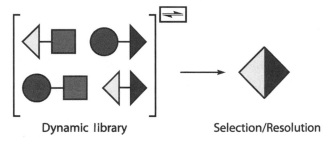

Dynamic library   Selection/Resolution

Figure 6.6 Internal dynamic combinatorial resolution (iDCR).

amplified products from reactions occurring in dynamic libraries. It also provides an alternative tool for controlling or halting dynamic combinatorial systems, and for demonstrating dynamics in biased equilibria.

6.3. DCR Systems: Examples

6.3.1. Resolution of Dynamic Combinatorial Thiolester Libraries [2,3]

Transthiolesterification represents the reaction between a thiolester and a thiol, a reversible reaction of fundamental biological importance [7]. For example, the thiolester acetyl-coenzyme A (Ac-CoA) is formed through a transthiolesterification during the oxidative decarboxylation of pyruvate in the Krebs cycle. The reaction normally proceeds rapidly and effectively in water, a feature that is highly desirable, but rarely applicable for organic transformations. This enables systems using biological targets and a possibility to work more closely with life sciences and pharmaceutical applications.

The dynamic features of the transthiolesterification reaction have been probed for a range of components of different character, resulting in potent thiolester libraries. The libraries were also exposed to selection by a variety of different hydrolases, where performance and selectivity of the dynamic resolution process were investigated. Following the catalytic action of the enzyme, the products would be expelled from the active site, thus rendering the site free to host more of DCL constituents and forcing the dynamic system to run to completion. Substrates to the biocatalyst would therefore be selectively produced and easily identified using this self-screening dynamic system. This process generates more of the best bound species by re-equilibration of the DCL.

Prior to these investigations, there had been no reported use of transthiolesterification in DCC systems. Since these reports, however, the potential of this process has been further explored in other protocols [8–11].

Target Proteins

Hydrolases catalyze the hydrolytic cleavage of C–O, C–N, C–C, and some other bonds including phosphoric anhydride bonds. They possess several attractive features, such as broad substrate selectivity and high stereospecificity. This has made them a popular choice for the conduction of many biotransformations as well as a powerful addition to the organic chemistry toolbox. Hydrolases also often catalyze several related reactions, such as condensations and alcoholysis.

The cholinesterases, acetylcholinesterase and butyrylcholinesterase, are serine hydrolase enzymes. The biological role of acetylcholinesterase (AChE, EC 3.1.1.7) is to hydrolyze the neurotransmitter acetylcholine (ACh) to acetate and choline (Scheme 6.1). This plays a role in impulse termination of transmissions at cholinergic synapses within the nervous system (Fig. 6.7) [12,13]. Butyrylcholinesterase (BChE, EC 3.1.1.8), on the other hand, has yet not been ascribed a function. It tolerates a large variety of esters and is more active with butyryl and propionyl choline than with acetyl choline [14]. Structure–activity relationship studies have shown that different steric restrictions in the acyl pockets of AChE and BChE cause the difference in their specificity with respect to the acyl moiety of the substrate [15]. AChE hydrolyzes ACh at a very high rate. The maximal rate for hydrolysis of ACh and its thio analog acetylthiocholine are around 10^9 M^{-1} s^{-1}, approaching the diffusion-controlled limit [16].

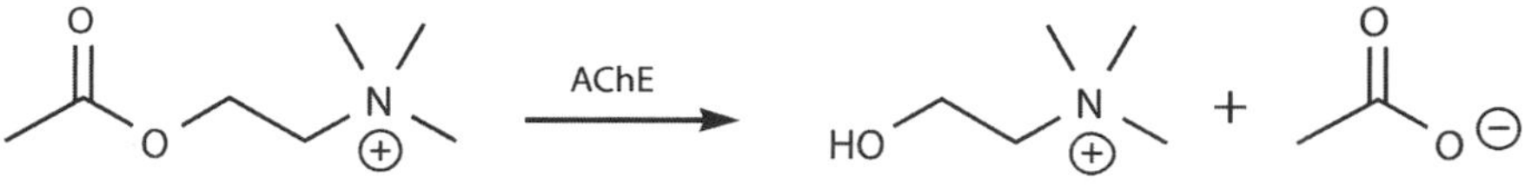

Scheme 6.1   The AChE catalyzed hydrolysis of ACh to choline and acetate.

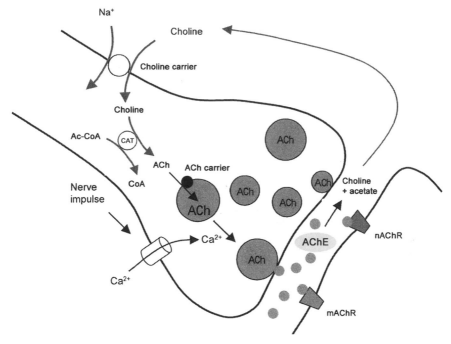

Figure 6.7   Schematic representation on the biological role of AChE in a cholinergic synapse.

Dynamics of Transthiolesterification

In order to explore the dynamic properties of transthiolesterification, a range of thiols (**1–13**) were investigated (Fig. 6.8). The selected thiols were of diverse nature and differed in respect to structure and properties. In this way, the potential for constructing large and diverse DCLs was investigated. Although all thiols displayed large structural variety, they were all chosen so as to be soluble in aqueous media at neutral pH and thereby be suitable for experiments under biological conditions.

The dynamic features of each of the thiols were subsequently evaluated in transthiolesterification reactions in buffered D_2O solution (NaOD/D_3PO_4, pD 7.0) with the ACh analog acetylthiocholine [ASCh (**14**), Table 6.1]. Formation/thiolysis of each thiolester was carefully followed by ^1H-NMR spectroscopy at different time intervals, and exchange rate and equilibrium composition were determined for each combination. The rate of exchange was directly correlated to the pK_a of the thiols; the lower the pK_a, the faster the exchange reaction (Table 6.1). Thiols having pK_a values lower than 8.5 reached equilibrium very rapidly. The results also showed that the majority of thiols produce equilibrium concentrations that are close to

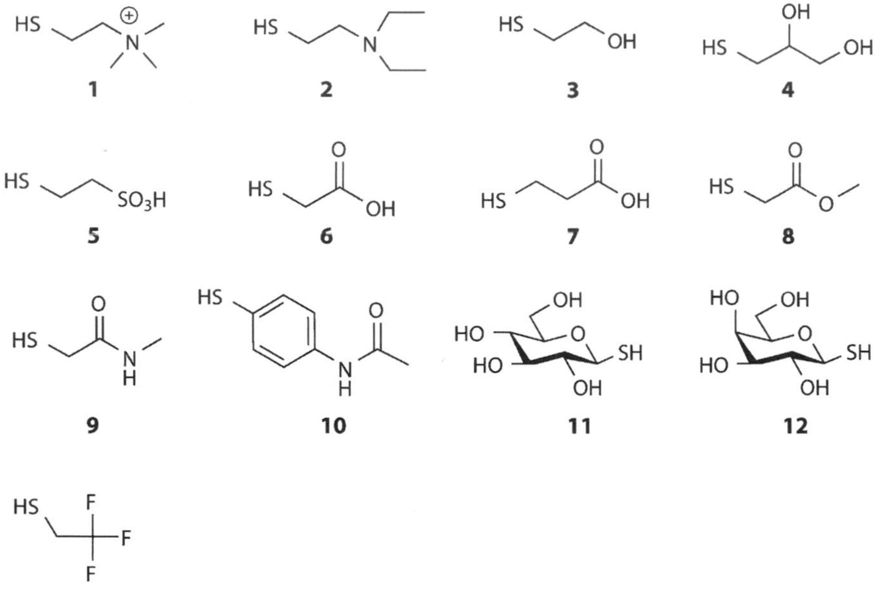

Figure 6.8 Selected thiols for investigating the dynamics of transthiolesterification.

Table 6.1 Individual Properties of Thiols (**1–13**) in Dynamic Transthiolesterification

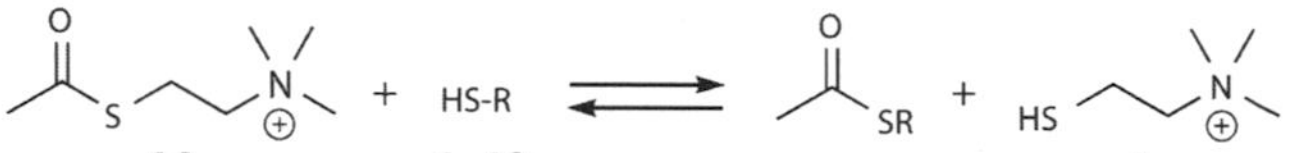

		Exchange with ASCh	
Thiol	**pK$_a$**	**Ratio**	**t½ (min)**
1	7.7	–	–
2	7.7	1.0:0.6	<15
3	9.7	1.0:1.2	40
4	9.5	1.0:1.7	55
5	9.1	1.0:0.9	25
6	9.8	1.0:1.0	40
7	10.3	1.0:1.0	40
8	7.8	1.0:1.0	<15
9	8.5	1.0:1.2	<15
10	6.1	1.0:0.1	<15
11	7.7	1.0:0.2	<15
12	7.7	1.0:0.2	<15
13	7.3	1.0:1.0	<15

the concentration of acetylthiocholine, thus showing near-isoenergetic behavior. This is a feature that is desirable, but rarely seen in dynamic combinatorial systems. It prevents the formation of libraries that are strongly biased toward certain products, which may make it energetically costly to shift the equilibrium. Thiolesters from secondary thiols (**10–12**) were however considerably less stable compared to acetylthiocholine. For the aromatic thiol (**10**), and the 1-thio-β-D-glycopyranoses (**11** and **12**), the ratio was clearly shifted in favor of the reactants, and only 10–20% of the thiols were present as the corresponding thiolesters at equilibrium. When comparing all thiols, these components showed the largest differences in reactivity.

The dynamic properties of thiolesters with a variety of acyl groups (**16–22**), including both linear (**16–20**) and branched (**21** and **22**) structures, were also investigated (Figs. 6.9 and 6.10). To keep the acyl compounds soluble at neutral pH, 3-sulfanylpropionic acid (**7**) was used as thiol counterpart. The kinetics was estimated from two different dynamic libraries DCL-A and DCL-B (Fig. 6.10). DCL-A consisted of the five linear thiolesters (**16–20**) and thiocholine (**1**), while in DCL-B the longer linear components (**19** and **20**) were exchanged for branched ones (**21** and **22**). The exchange was monitored by ¹H-NMR spectroscopy at pD 7.0 and room temperature. As expected, the results from these libraries implied that the branched acyl groups (**21** and **22**) reduced the exchange rate of the libraries due to the increased steric bulk surrounding the carbonyl functionality.

As mentioned above, in order to avoid creating severely biased libraries, components of comparable reactivity should be used in the library construction. Due to slow exchange rates or unfavorable equilibria, some thiols and acyl components were excluded from further studies. However,

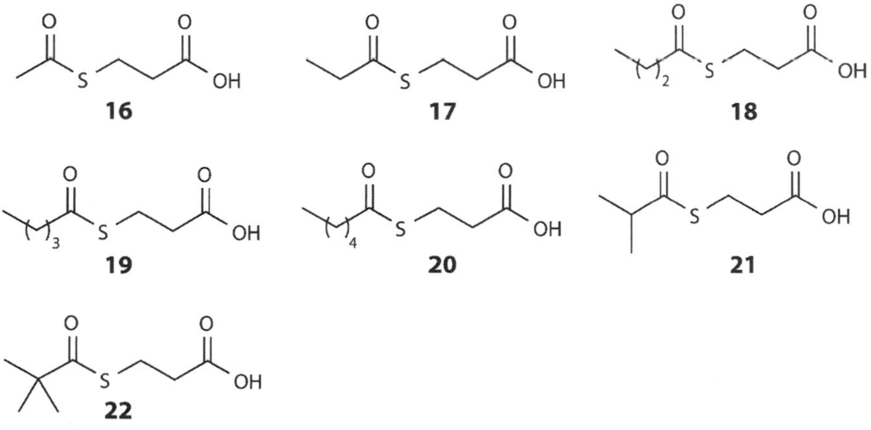

Figure 6.9 Acyl group variation for investigation of the dynamics of transthiolesterification.

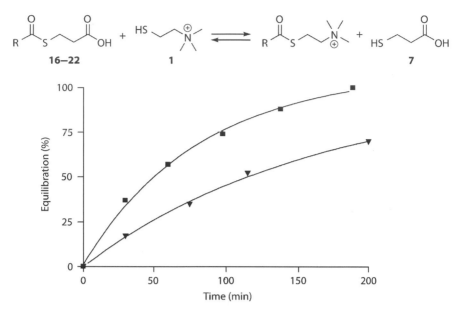

Figure 6.10 Equilibration of dynamic thiolester libraries: (■) DCL-A from thiol **1** and thiolesters **16–20**; (▼) DCL-B from thiol (**1**) and thiolesters **16–18, 21,** and **22** (modified from Reference 2).

they are not generally disqualified for use in thiolester libraries, but may well be part of dynamic libraries for other purposes. One way to increase the rate of transthiolesterification is to increase the pH of the solution, although one should be aware that an increase in basicity also accelerates the rate of unwanted hydrolysis.

DCR of Cholinesterase Substrates (Scheme 6.2) [2,3]

The DCLs were generated from a series of thiolesters and thiols, respectively. The reaction between five thiolesters (**16–20**) and four thiols (**1, 2, 5,** and **9**) generated library DCL-C (Scheme 6.3). For every thiol added, five additional thiolesters are formed, making a total of 25 thiolesters, all constantly undergoing exchange with the five thiols during the whole process.

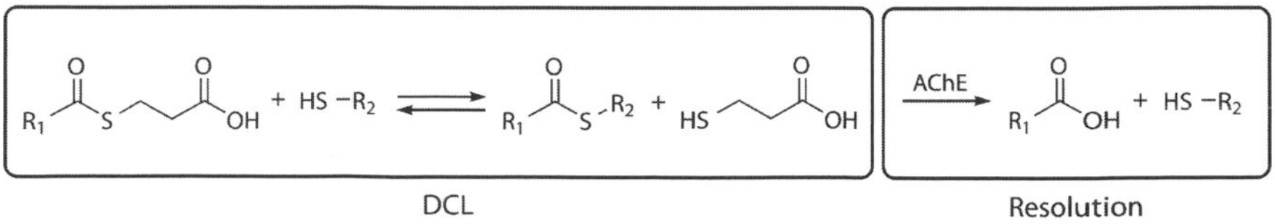

DCL   Resolution

Scheme 6.2 Resolution of cholinesterase substrates from a thiolester DCL.

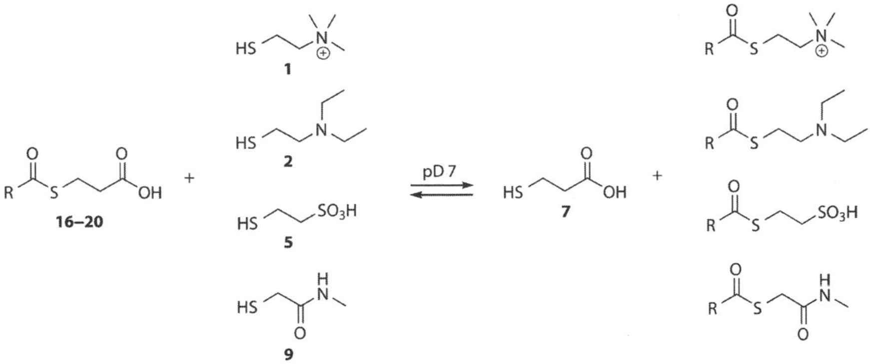

Scheme 6.3   DCL-C: A 25-compound DCL constructed for screening of hydrolase substrates.

The transthiolesterification reaction took place efficiently under mild conditions in aqueous media by simply mixing all the components. The library generation process was initiated by mixing equimolar amounts of all acyl components with five equivalents of each thiol. Since thiol component (**7**) is connected to the five different acyl functionalities at t_0, this ensured equal quantities of all thiol components in the system. Thus, the resulting concentrations of the formed constituents were relatively comparable, and the library showed close to isoenergetic behavior.

Treatment of the thiolester library with acetylcholinesterase resulted in the best substrate being immediately recognized by the enzyme and subsequently hydrolyzed (Fig. 6.11). This resulted in loss of the acyl component from the library, which forced the library to reconstitute to accommodate the increasing amounts of free thiol and to generate more of the hydrolyzed species. Over time, two of the acyl functionalities, the acetyl and propionyl groups, proved to be mainly acted upon by the enzyme, with the acetyl species being more rapidly hydrolyzed than the propionyl counterpart. Only after substantial hydrolysis of the two main substrates did the enzyme start hydrolyzing the butyrate ester, and even then, the reaction was considerably slower. This substantial lag phase may be caused by inhibitory activities of the present esters [17]. The other acyl groups remained untouched by the enzyme, a result which is in accordance with the known specificity of acetylcholinesterase.

To test the selectivity of the self-screening process, six other enzymes belonging to the hydrolase family were tested under the same set of conditions as in DCL-A. These enzymes were butyrylcholinesterase, horse liver esterase (HLE, EC 3.1.1.1), *Candida cylindracea* lipase (CCL, EC 3.1.1.3), β-galactosidase (β-Gal, EC 3.2.1.23), trypsin (EC 3.4.21.4), and

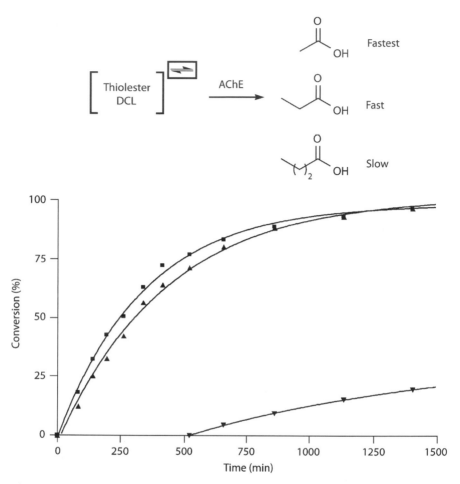

Figure 6.11 Formation of acetate (■), propionate (▲), and butyrate (▼) hydrolysis products in DCL-C (modified from Reference 2).

subtilisin Carlsberg (Sub C, EC 3.4.21.62). The dynamic libraries were thus exposed to each of the enzymes and the formation of hydrolysis products analyzed (Table 6.2). All hydrolases acting on carboxylic ester bonds (EC 3.1.1.X) showed some activity, although CCL acted very modestly, most likely due to the somewhat unfavorable medium for this enzyme. In contrast to acetylcholinesterase, butyrylcholinesterase acted on all acyl groups and hydrolyzed all groups in roughly the same time. This result is well in accordance with the known substrate pattern for this enzyme. The esterase from horse liver (HLE) showed a pattern in which the longer acyl chains were slightly faster than their shorter counterparts. For the two proteases trypsin and subtilisin, only the latter shows some activity under

Table 6.2   Acyl Product Formation for Seven Hydrolases and BSA with DCL-A

Enzyme	Product yield (%)[a]				
	Acetate	**Propionate**	**Butyrate**	**Valerate**	**Caproate**
AchE	50	45	–	–	–
BChE	37	42	44	44	43
HLE	16	19	20	23	31
CCL	–	<5	<5	<5	<5
β-Gal	–	–	–	–	–
Trypsin	–	–	–	–	–
Sub C	–	<5	<5	9	14
BSA	–	–	–	–	–

[a]$t = 210$ min ($t_{1/2}$ for acetate/AChE)

these conditions, also with some selectivity for the longer acyl chains. The hydrolase belonging to the glycolases, β-galactosidase (β-Gal), did not show any activity, as did bovine serum albumin (BSA), used as control.

6.3.2. Resolution of Dynamic Combinatorial Cyanohydrin Libraries [4]

The formation of carbon–carbon bonds has always been one of the key challenges in synthetic organic chemistry, and particularly methods to obtain optically pure products are of fundamental importance. In DCC, however, with the exception of the powerful alkene metathesis reaction, C–C bond formation has only been explored in a few systems [1,5,6,18–20].

The reactions between cyanide ion and ketones or aldehydes, generating cyanohydrin compounds, were first discovered in 1872 [21]. These classical reactions represent an important route to C–C bond formation, and the resulting cyanohydrins are versatile building blocks that can be transformed into a variety of functional groups. For example, these structures can undergo fluorination, hydrolysis, and reduction to provide α-fluoronitriles, α-hydroxyketones, α-hydroxycarboxylic acids, and β-hydroxyamines [22–25]. The reactions are also amenable to stereocontrol, and asymmetric cyanohydrin synthesis has been widely explored using, for example, asymmetric catalysis and enzymatic resolution [26–29]. By use of chemoenzymatic protocols optically active cyanohydrins can, for example, be produced in high yield and high enantiomeric excess (ee) [29].

These features, together with the reversible, thermoneutral nature of the cyanohydrin formation, make the reaction potentially useful for applications in dynamic chemistry. This was also recently shown in a DCR study [4], where selective enzyme-mediated cyanohydrin transesterification could be used to resolve the dynamic systems and identify optimal constituents. The dynamics of the reversible C–C bond formation was thus used to generate cyanohydrin DCLs, and *in situ* enzymatic resolution yielded particular stereoenriched cyanoacetate products.

Target Proteins

Lipases (EC 3.1.1.3) are ubiquitous enzymes belonging to the esterase class of hydrolases (see earlier section) and are found in most living organisms. In nature, lipases catalyze the hydrolytic cleavage of triglycerides into fatty acids and glycerol, or into fatty acid and mono- or diglyceride, at a water–oil interface (Scheme 6.4).

Lipases have also been proven to recognize a broad range of unnatural substrates in either aqueous or nonaqueous phase to catalyze (trans)esterification reactions. In addition, lipases specifically recognize enantiomeric molecules or enantiotopic groups on prostereogenic molecules, resulting in asymmetric transformations of high stereoselectivity/stereospecificity. This significant property can be used to achieve kinetic resolution (KR) of racemic compounds or obtain quantitative asymmetrization of prochiral or *meso* compounds. Generally, lipases show higher enantioselectivity when resolving secondary alcohols compared to primary or tertiary alcohols. The reactivity pattern follows an empirical rule originally proposed by Kazlauskas and coworkers. According to this rule, the fast-reacting enantiomer has the configuration displayed in Fig. 6.12. This was later explained using X-ray studies and stereoelectronic theory [30–35]. Furthermore, lipases are often commercially available, require no cofactors, and are easily recoverable.

Dynamics of the Cyanohydrin Reaction

In order to generate the dynamic cyanohydrin systems, several cyanide sources can be used, for example, cyanide salts, TMSCN, and cyanohydrin adducts such as acetone cyanohydrin. The latter method represents a means to form cyanohydrin DCLs under mild conditions, where acetone cyanohydrin is treated with amine base to release the cyanide ion together with acetone in organic solvents. The resulting cyanide ion then reacts with the set of aldehydes (or ketones), giving rise to the corresponding cyanohydrin adducts

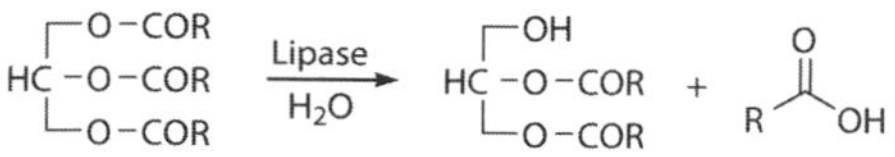

Scheme 6.4   The biological function of lipases.

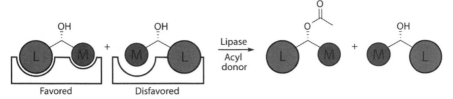

Figure 6.12 Empirical rule for predicting the preferred enantiomer in lipase-catalyzed acylation of a secondary alcohol. L represents the largest substituent and M the medium-sized group.

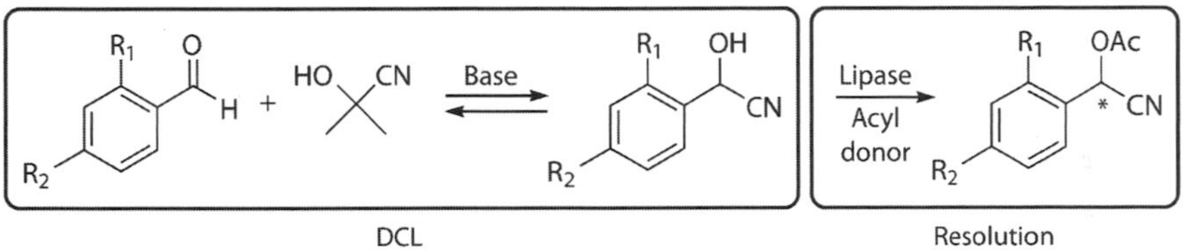

Scheme 6.5 Resolution of lipase substrates from a cyanohydrin DCL.

and establishing the dynamic system. The rate of the cyanohydrin reaction is generally fast, and can easily be further accelerated by varying the reaction conditions, for example, temperature, base, and solvents. In the present case, several bases and solvents were applied to the cyanohydrin reaction, and the temperature was varied to enhance the reaction rate. The optimized reaction conditions for generating the cyanohydrin DCLs were then selected as equimolar amounts of carbonyl compound, acetone cyanohydrin, and triethylamine in chloroform at room temperature.

DCR of Lipase Substrates (Scheme 6.5) [4]

The dynamic cyanohydrin libraries were generated from equimolar amounts of five different benzaldehydes (**23–27**) and acetone cyanohydrin (**28**) in the presence of triethylamine. This resulted in release of cyanide ion and rapidly provided the cyanohydrin adducts (**29–33**), all as racemic mixtures of both enantiomers (DCL-D, Scheme 6.6).

The benzaldehydes (**23–27**) were chosen in order to attain similar reactivities in the cyanohydrin reactions, and thus securing sufficiently unbiased, isoenergetic dynamic systems. The rates of the individual reactions of these benzaldehydes were further investigated by ^{1}H-NMR, indicating the establishment of individual aldehyde–cyanohydrin equilibria within 30 minutes. Under these conditions, the complete dynamic cyanohydrin system generated from cyanide and benzaldehydes (**23–27**) then reached

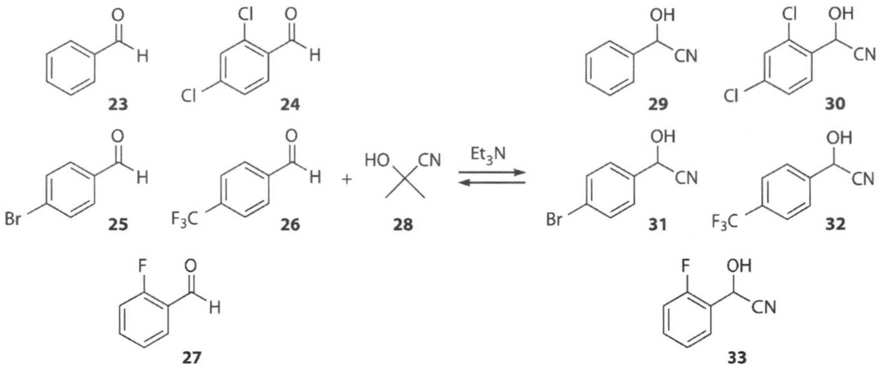

Scheme 6.6   DCL-D: Five benzaldehyde derivatives (**23–27**) and acetone cyanohydrine (**28**), forming cyanohydrin adducts (**29–33**).

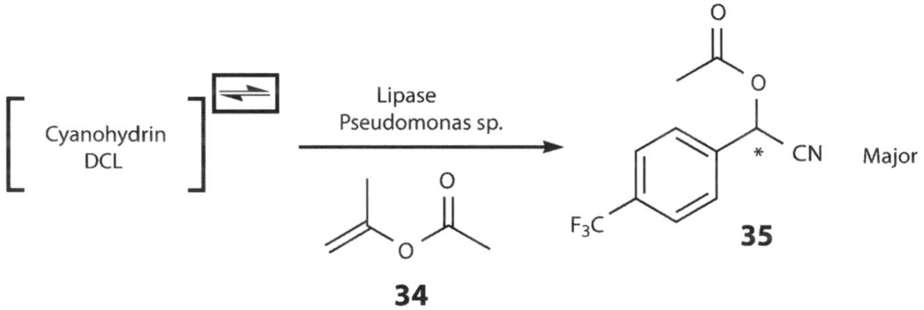

Scheme 6.7   Lipase-catalyzed resolution of cyanohydrin library DCL-D, yielding ester (**35**) as the major product.

equilibrium within 3 hours. The resulting cyanohydrin adducts (**29–33**) were however fully stable for longer times.

In conjunction with the establishment of the cyanohydrin DCL, the DCR process was subsequently addressed. Thus, selected lipases and a suitable acyl donor [isopropenyl acetate (**34**)] were applied to the system (Scheme 6.7). This selective enzymatic resolution of the DCL provided cyanoacetate product (**35**) as the major product at the reaction conditions used, thus demonstrating the efficiency of the concept.

6.3.3. Resolution of Dynamic Combinatorial Nitroaldol Libraries [1,5,6]

The nucleophilic addition of a nitroalkane to a carbonyl group, referred to as the nitroaldol or Henry reaction, is a very powerful C–C bond-forming reaction of great tradition and with numerous applications in synthetic organic chemistry [36–39]. Moreover, the diversity of further transformations of the nitroaldol adduct, including oxidation, reduction, and reductive nitration, provide efficient access to functionalized structural motifs, such as 1,2-amino alcohols, common in chemical and biological expressions. Recently, it has also been developed as an efficient C–C bond-forming route to DCLs [1,5,6]. These DCLs, generated under thermodynamic control, were then coupled in a kinetically controlled one-pot DCR process using lipase-mediated transesterification [1]. This asymmetric resolution of the DCLs by the enzyme led to the isolation of selected, and enantiomerically pure, β-nitroacctates in high yield.

Dynamic combinatorial nitroaldol libraries were also used to illustrate iDCR [5,6]. In this case, one of the library components was selected for its possibility to undergo an irreversible tandem cyclization reaction following equilibrium formation. This provided an internal kinetic selection pressure on the library, subsequently forcing the library to complete amplification of this novel reaction product. Furthermore, interesting crystalline properties were observed for one of the diastereoisomers of this isoindolinone-type product, providing a route to demonstrate consecutive resolution processes resolving coupled DCLs in a one-pot experiment.

Dynamics of the Nitroaldol Reaction

The nitroaldol reaction, similar to other aldol-type reactions, is thermodynamically controlled. It is very useful for a variety of aldehydes, but normally requires control of the equilibrium displacement to obtain useful yields. In these studies, this was however used as an advantage where β-nitroalcohol DCLs were efficiently formed from the reversible nitroaldol formation. The reaction can also be accelerated by the addition of different reagents, such as organic and inorganic bases, quaternary ammonium salts, or by use of ionic liquids. In order to optimize the nitroaldol reaction conditions for the DCR process, several bases were initially screened to find suitable conditions for DCL generation. The reaction was conducted in the presence of one equivalent each of nitroalkane, aldehyde, and base, and monitored by ^{1}H-NMR. By comparing the signals of the aldehyde and nitroalcohol adduct, the rate of the reaction was determined. Due to rapid and stable equilibration and, very importantly, high compatibility with subsequent enzymatic reactions, triethylamine proved to be optimal for use in this system.

DCR of Lipase Substrates (Scheme 6.8) [1]

The DCL was generated from equimolar amounts of five different aromatic aldehydes (**24, 26, 27, 36**, and **37**) and 2-nitropropane (**38**) to provide 10 β-nitroalcohol substrates including all enantiomers (DCL-E, Scheme 6.9).

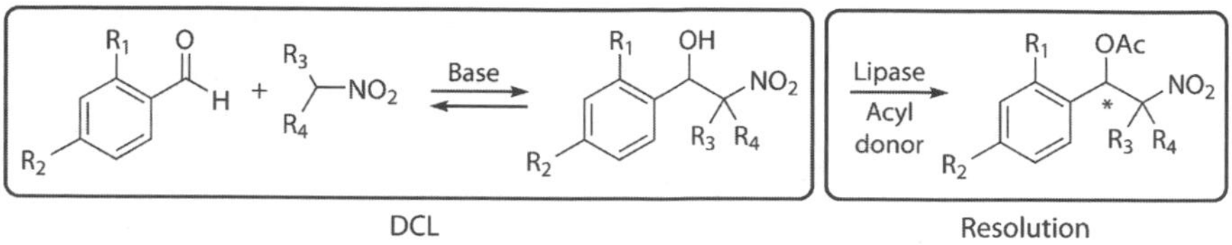

Scheme 6.8 Resolution of lipase substrates from a nitroaldol DCL.

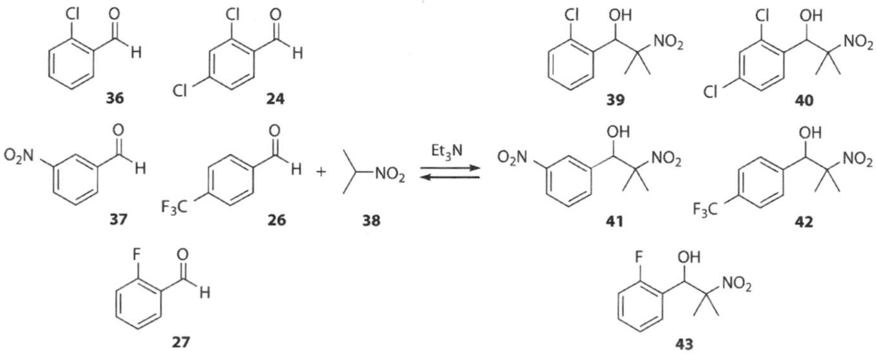

Scheme 6.9 DCL-E: Five benzaldehyde derivatives (**24, 26, 27, 36**, and **37**) and 2-nitropropane (**38**), forming nitroaldol adducts (**39–43**).

Although catalytic amounts of base proved sufficient for library generation, 10 equivalents were used in order to achieve a reasonable equilibration rate for the slightly different ratio between the quantities of aldehydes and nitroalkane (5:1). Furthermore, these benzaldehydes were chosen in view of their similar individual reactivity in the nitroaldol reaction, resulting in close to isoenergetic behavior in the produced DCL.

Figure 6.13 displays ^{1}H-NMR analysis of the resulting β-nitroalcohol DCL. After mixing benzaldehydes (**24, 26, 27, 36**, and **37**) with 2-nitropropane (**38**), equilibration was initiated by the addition of triethylamine. To allow faster equilibration, the exchange took place at 40°C, and all adducts were clearly present at equilibrium that was established after 18 hours. Equilibration also worked well at ambient temperature, albeit showing slower rates. Temperatures above 40°C, however, had a negative influence on the enantioselectivity of the subsequent enzymatic reaction and also caused slow decomposition of the β-nitroalcohol substrates.

Lipase-catalyzed transesterification of β-nitroalcohol substrates had not previously been reported and required careful optimization of the reaction conditions. A series of enzymes were screened, followed by acyl donors. From these results, the lipase *Pseudomonas cepacia* (PS-C I) (for more

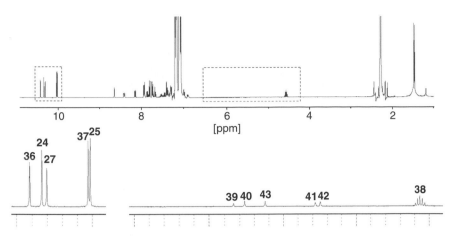

Figure 6.13 ¹H-NMR of nitroaldol DCL-E at equilibrium (modified from Reference 1).

information on lipases, see earlier section) and *p*-chlorophenyl acetate were selected as lipase and acyl donor, respectively, for the DCR system.

The PS-C I lipase was added together with five equivalents of *p*-chlorophenyl acetate (**44**) to the nitroaldol library at 40°C. This resulted in selective transesterification of two of the β-nitroalcohols, affording the corresponding acylated products (**45** and **46**; Fig. 6.14). The major product proved to be the ester **45**, produced from 3-nitrobenzaldehyde (**37**) and 2-nitropropane (**38**). Notably, the nitroaldol adduct **41**, serving as a substrate for the lipase reaction, was one of the least formed adducts in absence of enzyme. Nevertheless, this product was mainly selected in this process. A minor amount of the ester **46** was also formed, from 2-fluorobenzaldehyde (**27**) and 2-nitropropane (**38**), in the reaction. The overall yield for the two final products amounted to 24% after 24 hours, although resolution of the two products could be identified much earlier. Prolonged reaction times, however, resulted in improved reaction yields, reaching 80% after 14 days and almost completion (95%) after 20 days. The amplification effect was slightly improved when performing the reaction at ambient temperature and/ or using smaller amounts of enzyme. This, however, also increased the reaction time (65% yield in 14 days).

The nitroaldol-lipase DCR process could not only amplify specific β-nitroalcohol derivatives, but also lead to their asymmetric discrimination. HPLC analysis proved that the enantioselectivity of the process is very high, resulting in products of very high optical purity. The *R*-enantiomer of the ester **45** was resolved to 99% ee, and the *R*-enantiomer of the ester **46** to 98% ee.

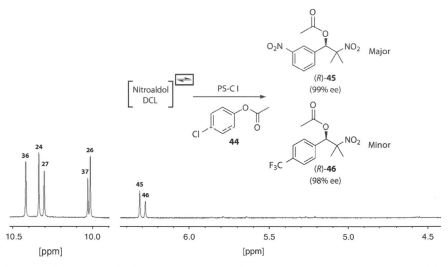

Figure 6.14   DCL-E in presence of PS-C I and *p*-chlorophenyl acetate (44) (modified from Reference 1).

Internal DCR of Nitroaldol Libraries (Scheme 6.10) [5,6]

iDCR was demonstrated by using a conceptual nitroaldol library including five benzaldehyde derivatives (**24, 36**, and **47–49**) and one nitroalkane (**50**, DCL-F, Scheme 6.11). The benzaldehydes, all with a unique substitution pattern, were selected in order to make analysis clear and simple. However, one of the benzaldehydes contained a cyano functionality in the 2-position (**49**), deliberately making it a candidate for subsequent tandem cyclization following nitroalcohol formation. 5-*exo-dig* type cyclizations of hydroxynitriles to the corresponding iminolactones are expected [40,41], albeit unexplored [42–45], intramolecular transformations, which in this case could lead to possible kinetic resolution of the library.

Library generation was initially demonstrated using one equivalent each of aldehydes (**24**, **36**, **47**, and **48**), excluding the cyanosubstituted candidate (**49**), together with one equivalent of nitroethane (**50**) in *d*-acetonitrile (DCL-G).

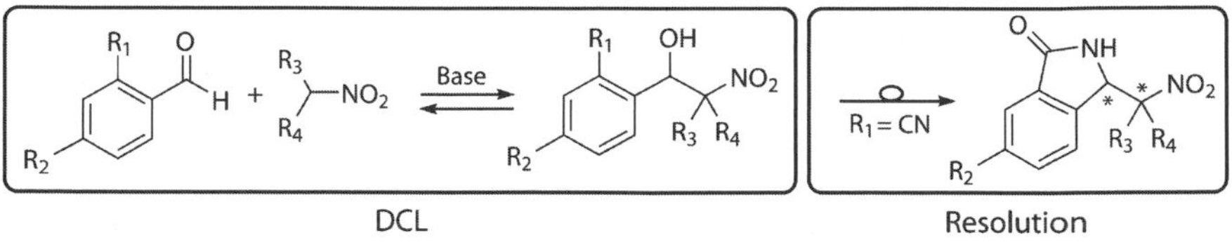

Scheme 6.10   Internal resolution of a nitroaldol DCL.

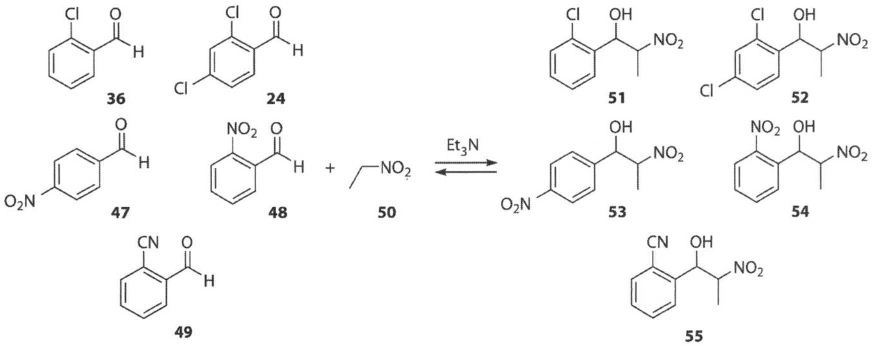

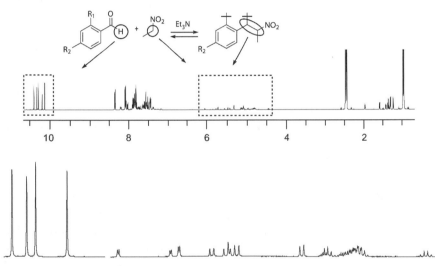

Scheme 6.11 DCL-F: Five benzaldehyde derivatives (**24, 36**, and **47–49**) and nitroethane (**50**), forming nitroaldol adducts (**51–55**).

Figure 6.15 ¹H-NMR of nitroaldol DCL-G at equilibrium (modified from Reference 6).

To initiate the equilibrium, triethylamine was added and the reaction followed frequently by ¹H-NMR. Under these conditions, equilibrium was reached within 3 hours (Fig. 6.15). Following this, the full library (DCL-F) including aldehyde (**49**) was constructed. This library initially behaved similarly to DCL-G, with several nitroaldoladducts forming competitively. A total of 20 individual β-nitroalcohol adducts is possible including enantiomers and diastereoisomers. However, within 1 hour, a clear amplification of a single diastereomeric pair, supposedly iminolactone (**56**), became clearly visible. The amplification effect then gradually continued until all

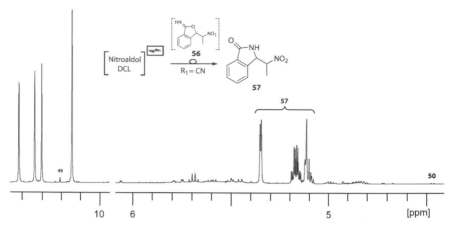

Figure 6.16   DCL-F after full resolution of compound **57** (modified from Reference 6).

previously formed nitroalcohols, 2-cyanobenzaldehyde, and nitroethane were consumed (Fig. 6.16).

An interesting observation was made upon isolation and characterization of the amplified product. NMR spectroscopic data together with supporting X-ray diffraction analysis clearly proved the compound not to be iminolactone (**56**), but rather lactam (**57**). Further mechanistic studies are yet to be carried out for the proposed rearrangement, but a few related systems are reported [46–48].

Detailed time-dependent NMR studies were also performed in order to evaluate the kinetic profile of the tandem reaction (Fig. 6.17). Experimental data was coherent with a reversibly–irreversibly-coupled model, suggesting the proposed rearrangement step to be faster compared to the iminolactone cyclization. The studies also revealed the forward nitroaldol formation to be of comparable rate to tandem cyclization, while the reverse nitroaldol reaction was slower and thereby rate determining for the overall DCR process.

Further studies of the isolated isoindolinone structure (lactam **57**) showed that one of the diastereoisomers had interesting crystalline properties. This was used in order to demonstrate how a DCR process could further be directed and diastereomerically amplified through a selection pressure caused by a phase change. Systems where DCLs are driven by crystallization have previously been reported [50–52], and have recently received new attention [53–56].

A sublibrary of the DCL-F system was used, including aldehydes (**36, 48**, and **49**) (DCL-G). When employing dilute conditions in *d*-chloroform,

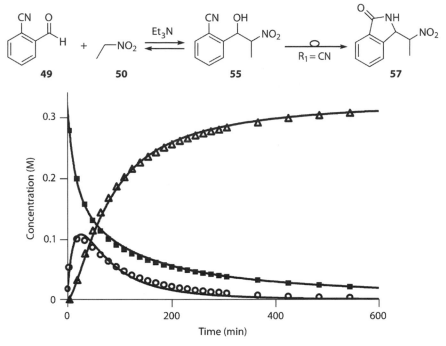

Figure 6.17 Reaction composition followed over time: (■) aldehyde (**49**), (○) nitroalcohol intermediate (**55**), and (∆) product (**57**) (modified from Reference 6). Lines represent fitted data from kinetic model [49].

an expected amplification and complete resolution of isoindolinone (**57**) (both diastereoisomers) was seen (Fig. 6.18a). These isomers formally also constitute a DCL, due to the presence of triethylamine, in the presence of which they are rapidly interconverted. When the concentration of the library was increased, however, a precipitate started to form in the reaction vessel (Fig. 6.18b). Powder and X-ray diffraction studies proved the precipitate to be close to the diastereomerically pure form (97:3) of the (*R*,*R*)/(*S*,*S*) isomer of **57**. Further optimization of precipitation conditions resulted in isolation of 63% overall yield of the single isomer from a mixed chloroform/hexane system. Important to note is that the isolated and amplified diastereoisomer is the opposite isomer of the one being more stable in solution (Fig. 6.18a), thus further increasing the amplification factor. This is the first example of coupled DCLs with distinctly coupled resolution processes. Furthermore, the product itself is a synthetically interesting motif, present in various drug-like compounds and natural products [57–59]. It also is a precursor to the widely used 1,2-diamines [60–63].

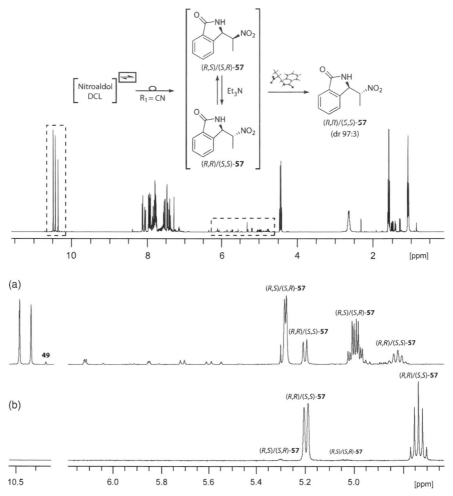

Figure 6.18 Selected ¹H-NMR spectra. (**a**) Diluted DCL-G after resolution. (**b**) Spectrum of the filtrated precipitate (modified from Reference 5).

6.4. Outlook and Conclusions

In this chapter, it has been demonstrated that DCR is an effective method for effective kinetic screening of dynamic libraries. By combining DCLs with enzyme-mediated reactions, complete resolution of the libraries could be achieved in a one-pot process. This approach also enables screening of complex DCLs without the necessity of equimolar amounts of target molecules.

It has furthermore been demonstrated that transthiolesterification is a highly useful way to generate DCLs under mild conditions in aqueous media, where the resulting system can be subjected to DCR in the presence of hydrolase enzymes. Biocatalysts can be efficiently used for targeting and rapidly identifying the best substrates formed in the DCL, where the outcome of the DCR system relies on the selectivity of the specific enzyme used. The cyanohydrin and nitroaldol (Henry) reactions have also been identified as new and efficient C–C bond-forming routes to DCL formation. The DCR processes were, in these cases, used to generate a collection of potential enzyme substrates, and to identify the best substrates for different lipases. One process also enabled the formation of enantiomerically pure β-nitroalcohol acetates in high yield.

Although DCR is an ideal concept for specific substrate analog searches, it is, however, not restricted to enzyme catalysis; it may be extended to include any secondary, selective synthetic/catalytic system, including organic and inorganic catalysts, and may be employed to rapidly screen reactions of catalysts for new substrates. An example of a broadening subconcept is internal DCR, where subsequent pressure from an intramolecular tandem reaction kinetically resolved a nitroaldol DCL. This is an interesting concept with possible use as a control element in DCC applications or a starting point for expanding irreversibly coupled DCC in other directions. One such interesting area is a possible use of DCC in systematized reaction discovery processes, letting unexpected, irreversible tandem reactions take place in DCLs, causing amplification of certain reaction products, which later could be analyzed and identified.

Looking ahead, there is definitely a promising future for DCC in general, and DCR in particular. During its short existence, DCR has made strong progress and developed into a more versatile and useful concept. More tolerable and solid reversible reaction systems have been demonstrated, which bodes well for future applications. Water tolerance is also an important feature that has been demonstrated. This, together with the low demands on selector quantities, is attractive, clearly enhancing the chances for making further inroads into the pharmaceutical industry. The expansion of the concept toward other applications, such as reaction discovery, is also an important development. Although still being in an early phase, it displays the scope and potential of this highly interesting concept in future.

References

1. Vongvilai, P.; Angelin, M.; Larsson, R.; Ramström, O. Dynamic combinatorial resolution: Direct asymmetric lipase-mediated screening of a dynamic nitroaldol library. *Angew. Chem. Int. Ed.* **2007**, *46*, 948–950.

2. Larsson, R.; Ramström, O. Dynamic combinatorial thiolester libraries for efficient catalytic self-screening of hydrolase substrates. *Eur. J. Org. Chem.* **2005**, 285–291.

3. Larsson, R.; Pei, Z.; Ramström, O. Catalytic self-screening of cholinesterase substrates from a dynamic combinatorial thioester library. *Angew. Chem. Int. Ed.* **2004**, *43*, 3716–3718.

4. Sakulsombat, M.; Vongvilai, P.; Ramström, O. *Manuscript* **2008**.

5. Angelin, M.; Fischer, A.; Ramström, O. Crystallization-induced secondary selection from a tandem driven dynamic combinatorial resolution process. *J. Org. Chem.* **2008**, *73*, 3693–3595.

6. Angelin, M.; Vongvilai, P.; Fischer, A.; Ramström, O. Tandem driven dynamic combinatorial resolution via Henry-iminolactone rearrangement. *Chem. Commun.* **2008**, *6*, 768–770.

7. Kielley, W. W.; Bradley, L. B. Glutathione thiolesterase. *J. Biol. Chem.* **1954**, *206*, 327–333.

8. Hadley, E. B.; Gellman, S. H. An antiparallel a-helical coiled-coil model system for rapid assessment of side-chain recognition at the hydrophobic interface. *J. Am. Chem. Soc.* **2006**, *128*, 16444–16445.

9. Woll, M. G.; Hadley, E. B.; Mecozzi, S.; Gellman, S. H. Stabilizing and destabilizing effects of phenylalanine → F5-phenylalanine mutations on the folding of a small protein. *J. Am. Chem. Soc.* **2006**, *128*, 15932–15933.

10. Leclaire, J.; Vial, L.; Otto, S.; Sanders, J. K. M. Expanding diversity in dynamic combinatorial libraries: Simultaneous exchange of disulfide and thioester linkages. *Chem. Commun.* **2005**, *15*, 1959–1961.

11. Woll, M. G.; Gellman, S. H. Backbone thioester exchange: A new approach to evaluating higher order structural stability in polypeptides. *J. Am. Chem. Soc.* **2004**, *126*, 11172–11174.

12. Sussman, J. L.; Harel, M.; Frolow, F.; Oefner, C.; Goldman, A.; Toker, L.; Silman, I. Atomic structure of acetylcholinesterase from *Torpedo californica*: A prototypic acetylcholine-binding protein. *Science* **1991**, *253*, 872–879.

13. Schumacher, M.; Camp, S.; Maulet, Y.; Newton, M.; MacPhee-Quigley, K.; Taylor, S. S.; Friedmann, T.; Taylor, P. Primary structure of *Torpedo californica* acetylcholinesterase deduced from its cDNA sequence. *Nature* **1986**, *319*, 407–409.

14. Main, A. R.; Soucie, W. G.; Buxton, I. L.; Arinc, E. The purification of cholinesterase from horse serum. *Biochem. J.* **1974**, *143*, 733–744.

15. Järv, J. Stereochemical aspects of cholinesterase catalysis. *Bioorg. Chem.* **1984**, *12*, 259–278.

16. Nolte, H. J.; Rosenberry, T. L.; Neumann, E. Effective charge on acetylcholinesterase active sites determined from the ionic strength dependence of association rate constants with cationic ligands. *Biochemistry* **1980**, *19*, 3705–3711.

17. Cho, S. J.; Garsia, M. L.; Bier, J.; Tropsha, A. Structure-based alignment and comparative molecular field analysis of acetylcholinesterase inhibitors. *J. Med. Chem.* **1996**, *39*, 5064–5071.

18. Boul, P. J.; Reutenauer, P.; Lehn, J.-M. Reversible Diels–Alder reactions for the generation of dynamic combinatorial libraries. *Org. Lett.* **2005**, *7*, 15–18.

19. Lins, R. J.; Flitsch, S. L.; Turner, N. J.; Irving, E.; Brown, S. A. Generation of a dynamic combinatorial library using sialic acid aldolase and in situ screening against wheat germ agglutinin. *Tetrahedron* **2004**, *60*, 771–780.

20. Lins, R. J.; Flitsch, S. L.; Turner, N. J.; Irving, E.; Brown, S. A. Enzymatic generation and in situ screening of a dynamic combinatorial library of sialic acid analogues. *Angew. Chem. Int. Ed.* **2002**, *41*, 3405–3407.

21. Urech, F. Über einige cyanderivate des acetons. *Ann. Chem. Pharm.* **1872**, *164*, 255–279.

22. Effenberger, F.; Gutterer, B.; Jäger, J. Stereoselective synthesis of (1R)- and (1R,2S)-1-aryl-2-alkylamino alcohols from (R)-cyanohydrins. *Tetrahedron: Asymmetry* **1997**, *8*, 459–467.

23. Stelzer, U.; Effenberger, F. Preparation of (S)-2-fluoronitriles. *Tetrahedron: Asymmetry* **1993**, *4*, 161–164.

24. Brussee, J.; Roos, E. C.; Van der Gen, A. Bioorganic synthesis of optically active cyanohydrins and acyloins. *Tetrahedron Lett.* **1988**, *29*, 4485–4488.

25. Krieble, V. L.; Wieland, W. A. Properties of oxynitrilase. *J. Am. Chem. Soc.* **1921**, *43*, 164–175.

26. Purkarthofer, T.; Skranc, W.; Schuster, C.; Griengl, H. Potential and capabilities of hydroxynitrile lyases as biocatalysts in the chemical industry. *Appl. Microbiol. Biotechnol.* **2007**, *76*, 309–320.

27. Lundgren, S.; Wingstrand, E.; Penhoat, M.; Moberg, C. Dual Lewis acid–Lewis base activation in enantioselective cyanation of aldehydes using acetyl cyanide and cyanoformate as cyanide sources. *J. Am. Chem. Soc.* **2005**, *127*, 11592–11593.

28. Hatano, M.; Ikeno, T.; Miyamoto, T.; Ishihara, K. Chiral lithium binaphtholate aqua complex as a highly effective asymmetric catalyst for cyanohydrin synthesis. *J. Am. Chem. Soc.* **2005**, *127*, 10776–10777.

29. Inagaki, M.; Hiratake, J.; Nishioka, T.; Oda, J. One-pot synthesis of optically active cyanohydrin acetates from aldehydes via lipase-catalyzed kinetic resolution coupled with in situ formation and racemization of cyanohydrins. *J. Org. Chem.* **1992**, *57*, 5643–5649.

30. Ema, T.; Kobayashi, J.; Maeno, S.; Sakai, T.; Utaka, M. Origin of the enantioselectivity of lipases explained by a stereo-sensing mechanism operative at the transition state. *Bull. Chem. Soc. Jpn.* **1998**, *71*, 443–453.

31. Cygler, M.; Grochulski, P.; Kazlauskas, R. J.; Schrag, J. D.; Bouthillier, F.; Rubin, B.; Serreqi, A. N.; Gupta, A. K. A structural basis for the chiral preferences of lipases. *J. Am. Chem. Soc.* **1994**, *116*, 3180–3186.

32. Burgess, K.; Jennings, L. D. Enantioselective esterifications of unsaturated alcohols mediated by a lipase prepared from *Pseudomonas sp. J. Am. Chem. Soc.* **1991**, *113*, 6129–6139.

33. Janssen, A. J. M.; Klunder, A. J. H.; Zwanenburg, B. Resolution of secondary alcohols by enzyme-catalyzed transesterification in alkyl carboxylates as the solvent. *Tetrahedron* **1991**, *47*, 7645–7662.
34. Xie, Z. F. Pseudomonas fluorescens lipase in asymmetric synthesis. *Tetrahedron: Asymmetry* **1991**, *2*, 733–750.
35. Kazlauskas, R. J.; Weissfloch, A. N. E.; Rappaport, A. T.; Cuccia, L. A. A rule to predict which enantiomer of a secondary alcohol reacts faster in reactions catalyzed by cholesterol esterase, lipase from *Pseudomonas cepacia*, and lipase from *Candida rugosa*. *J. Org. Chem.* **1991**, *56*, 2656–2665.
36. Purkarthofer, T.; Gruber, K.; Gruber-Khadjawi, M.; Waich, K.; Skranc, W.; Mink, D.; Griengl, H. A biocatalytic Henry reaction – The hydroxynitrile lyase from *Hevea brasiliensis* also catalyzes nitroaldol reactions. *Angew. Chem. Int. Ed.* **2006**, *45*, 3454–3456.
37. Palomo, C.; Oiarbide, M.; Mielgo, A. Unveiling reliable catalysts for the asymmetric nitroaldol (Henry) reaction. *Angew. Chem. Int. Ed.* **2004**, *43*, 5442–5444.
38. Luzzio, F. A. The Henry reaction: Recent examples. *Tetrahedron* **2001**, *57*, 915–945.
39. Henry, L. Formation synthétique d'alcools nitrés. *C. R. Hebd. Seances Acad. Sci.* **1895**, *120*, 1265–1268.
40. Baldwin, J. E.; Thomas, R. C.; Kruse, L. I.; Silberman, L. Rules for ring-closure: Ring formation by conjugate addition of oxygen nucleophiles. *J. Org. Chem.* **1977**, *42*, 3846–3852.
41. Baldwin, J. E. Rules for ring-closure. *J. Chem. Soc. Chem. Commun.* **1976**, *18*, 734–736.
42. Vasil'ev, A. N.; Lyshchikov, A. N.; Nasakin, O. E.; Kayukov, Y. S.; Tafeenko, V. A. Reduction of alkyl 2-amino-5,6-dialkyl-3-cyanopyridine-4-carboxylates. *Russ. J. Org. Chem.* **2005**, *41*, 279–282.
43. Langer, P.; Appel, B. Efficient synthesis of benzopyrano[2,3-b]pyridines by sequential reactions of 1,3-bis-silyl enol ethers with 3-cyanobenzopyrylium triflates. *Tetrahedron Lett.* **2003**, *44*, 5133–5135.
44. Rouillard, A.; Deslongchamps, P. Synthesis of a pentacyclic lactone related to quinovaic acid and emmolactone using an anionic polycyclization strategy. *Tetrahedron* **2002**, *58*, 6555–6560.
45. Benedetti, F.; Berti, F.; Fabrissin, S.; Gianferrara, T. Intramolecular ring opening of epoxides by bis-activated carbanions. The influence of ring size on reactivity and selectivity. *J. Org. Chem.* **1994**, *59*, 1518–1524.
46. Song, Y. S.; Lee, C. H.; Lee, K.-J. Application of Baylis–Hillman methodology in a new synthesis of 3-oxo-2,3-dihydro-1H-isoindoles. *J. Heterocycl. Chem.* **2003**, *40*, 939–941.
47. Sato, R.; Ohmori, M.; Kaitani, F.; Kurosawa, A.; Senzaki, T.; Goto, T.; Saito, M. Synthesis of isoindoles. Acid- or base-induced cyclization of 2-cyanobenzaldehyde with alcohols. *Bull. Chem. Soc. Jpn.* **1988**, *61*, 2481–2485.

48. Sato, R.; Senzaki, T.; Goto, T.; Saito, M. Novel synthesis of 3-(N-substituted amino)-1-isoindolenones from 2-cyanobenzaldehyde with amines. *Chem. Lett.* **1984**, *9*, 1599–1602.

49. Hoops, S.; Sahle, S.; Gauges, R.; Lee, C.; Pahle, J.; Simus, N.; Singhal, M.; Xu, L.; Mendes, P.; Kummer, U. COPASI – A COmplex PAthway SImulator. *Bioinformatics* **2006**, *22*, 3067–3074.

50. Chow, C.-F.; Fujii, S.; Lehn, J.-M. Crystallization-driven constitutional changes of dynamic polymers in response to neat/solution conditions. *Chem. Commun.* **2007**, *42*, 4363–4365.

51. Baxter, P. N. W.; Lehn, J.-M.; Kneisel, B. O.; Fenske, D. Self-assembly of a symmetric tetracopper box-grid with guest trapping in the solid state. *Chem. Commun.* **1997**, *22*, 2231–2232.

52. Baxter, P. N. W.; Lehn, J. M.; Rissanen, K. Generation of an equilibrating collection of circular inorganic copper(I) architectures and solid-state stabilization of the dicopper helicate component. *Chem. Commun.* **1997**, *14*, 1323–1324.

53. Hutin, M.; Cramer, C. J.; Gagliardi, L.; Shahi, A. R. M.; Bernardinelli, G.; Cerny, R.; Nitschke, J. R. Self-sorting chiral subcomponent rearrangement during crystallization. *J. Am. Chem. Soc.* **2007**, *129*, 8774–8780.

54. Haussmann, P. C.; Khan, S. I.; Stoddart, J. F. Equilibrating dynamic [2]rotax-anes. *J. Org. Chem.* **2007**, *72*, 6708–6713.

55. Pentecost, C. D.; Chichak, K. S.; Peters, A. J.; Cave, G. W. V.; Cantrill, S. J.; Stoddart, J. F. A molecular Solomon link. *Angew. Chem. Int. Ed.* **2007**, *46*, 218–222.

56. Dumitru, F.; Petit, E.; van der Lee, A.; Barboiu, M. Homo- and heteroduplex complexes containing terpyridine-type ligands and Zn^{2+}. *Eur. J. Inorg. Chem.* **2005**, *21*, 4255–4262.

57. Lamblin, M.; Couture, A.; Deniau, E.; Grandclaudon, P. A concise first total synthesis of narceine imide. *Org. Biomol. Chem.* **2007**, *5*, 1466–1471.

58. Hardcastle, I. R.; Ahmed, S. U.; Atkins, H.; Farnie, G.; Golding, B. T.; Griffin, R. J.; Guyenne, S.; Hutton, C.; Källblad, P.; Kemp, S. J.; Kitching, M. S.; Newell, D. R.; Norbedo, S.; Northen, J. S.; Reid, R. J.; Saravanan, K.; Willems, H. M. G.; Lunec, J. Small-molecule inhibitors of the MDM2-p53 protein–protein interaction based on an isoindolinone scaffold. *J. Med. Chem.* **2006**, *49*, 6209–6221.

59. Comins, D. L.; Schilling, S.; Zhang, Y. Asymmetric synthesis of 3-substituted isoindolinones: Application to the total synthesis of (+)-lennoxamine. *Org. Lett.* **2005**, *7*, 95–98.

60. Lowden, C. T.; Mendoza, J. S. Solution phase parallel synthesis of 1,2-phenethyldiamines. *Tetrahedron Lett.* **2002**, *43*, 979–982.

61. Lucet D.; Sabelle, S.; Kostelitz, O.; Le Gall, T.; Mioskowski, C. Enantioselective synthesis of a-amino acids and monosubstituted 1,2-diamines by

conjugate addition of 4-phenyl-2-oxazolidinone to nitro alkenes. *Eur. J. Org. Chem.* **1999**, 2583–2591.

62. Lucet, D.; Le Gall, T.; Mioskowski, C. The chemistry of vicinal diamines. *Angew. Chem. Int. Ed.* **1998**, *37*, 2580–2627.

63. Bennani, Y. L.; Hanessian, S. *trans*-1,2-Diaminocyclohexane derivatives as chiral reagents, scaffolds, and ligands for catalysis: Applications in asymmetric synthesis and molecular recognition. *Chem. Rev.* **1997**, *97*, 3161–3195.

Chapter 7

Dynamic Combinatorial Chemistry and Mass Spectrometry: A Combined Strategy for High Performance Lead Discovery

Sally-Ann Poulsen and Hoan Vu

7.1. Introduction

A "pipeline problem" has emerged within the pharma industry owing to a combination of declining approval rates for new drugs, expiration of patents on existing drugs, and alarming attrition metrics with 90% of new drug candidates exiting the pipeline prior to approval [1]. Drug lead discovery is a challenging front-end task of the drug discovery pipeline, and improved performance in lead discovery is without doubt the preferred trajectory for modern drug discovery campaigns. There is wide consensus that conceptually novel chemistry pursued within an academic setting could invigorate the lead discovery aspect of the pipeline [2]. Provided the chemistry program has a deliberate focus on drug discovery and engages with complementary expertise in biology and preclinical science (typically big pharma), it is well placed to facilitate and advance the discovery of lead compounds. Fragment-based drug discovery (FBDD) is a recent innovation that is now quite well established as a new tool for the pharma industry. FBDD protocols assess collections of "small" chemical structures, known as fragments (molecular weight $\sim< 250$ Da), in a screening campaign to identify those with affinity for the biological target. The typical affinity of a fragment is low, and hence follow-up lead optimization strategies are necessary. A key rationale for screening fragments versus collections of "drug-sized" molecules is that the optimization of drug-like properties that are necessary to transform the fragment lead to a candidate drug may benefit from a greater volume of chemical optimization space in which to operate when compared to already larger "drug-sized" molecules as leads. The commercial pressures and attrition metrics of the current pharma landscape have typically led to the adoption of Lipinski's rules as an early stage filter in drug optimization protocols [3] so that overall this FBDD strategy sits quite well within the existing pharma research framework. A comprehensive account of FBDD is outside the scope of this chapter; the field is however reasonably widespread with publications authored by those in both academia and the pharma industry—the interested reader is directed to recent reviews [4].

7.2. Target-Guided Medicinal Chemistry

Target-guided synthesis is readily conceptualized by the familiar "lock and key" host–guest descriptors of Emil Fischer, wherein the target acts as the host to facilitate the "correct" assembly of building blocks leading to synthesis of a complementary guest small molecule. This description encapsulates the principle of two complementary approaches in target-guided medicinal chemistry: (i) dynamic combinatorial chemistry (DCC) (linking fragments under thermodynamic control) and (ii) *in situ* click chemistry (linking fragments under kinetic control). In target-guided medicinal chemistry the target biomolecule assists the medicinal chemist to select and synthesize those reaction products that have a high affinity for the target from all potential reaction products accessible through the chemistry employed. In the context of protein targets, DCC was articulated in a proof-of-concept format by Huc and Lehn in 1997 [5] and with click chemistry by Sharpless, Finn, and coworkers in 2002 [6]. These novel synthetic approaches have since prospered as elegant and vibrant research disciplines in their own right, each with broadly scoped potential applications of which drug discovery is just one. DCC and click chemistry have evolved independent of, but alongside, FBDD. Both these target-guided strategies typically use fragments as building blocks; the fragments are covalently linked with the final product distribution determined by molecular recognition interactions with the target biomolecule. Target-guided synthesis, including both DCC and click chemistry, represents an extension of FBDD into the realms of target-guided fragment optimization.

7.2.1. Dynamic Combinatorial Chemistry

Performing synthetic chemistry to covalently link fragments under conditions that preserve the structural and functional integrity of a protein is challenging; however, this is now reasonably well established for DCC. The key feature of DCC is the reversible chemical reaction that mediates exchange of the building block fragments [7–9]. For drug discovery applications of DCC, the selection of linked fragments ideally occurs in the same environment as the equilibration reaction, and this commands that the reaction fulfill several requirements in addition to reversibility. The exchange reaction should (i) occur at a rate that allows equilibrium to be reached within an acceptable time; (ii) be bio-orthogonal with the template biomolecule, that is, with reactivity inert to the functional groups of the biomolecule; and (iii) operate under conditions that preserve the biologically relevant conformation of the biomolecule—typically aqueous buffer at physiological temperature and pH. Reaction conditions should not disrupt noncovalent interactions involved in molecular recognition between biomolecules and fragments; finally, the reactivity of all fragments should be similar to allow access to unbiased library compositions so as to ensure all possible constituents have the opportunity to interact with the target. The reversible reactions that meet these requirements have been detailed elsewhere in this book and in a recent review [7].

7.2.2. Click Chemistry

We have witnessed a tremendous volume of recent literature in relation to click chemistry. The premier transformation of click chemistry concerns the 1,3-dipolar cycloaddition reaction (1,3-DCR) of organic azides with terminal acetylenes to yield 1,2,3-triazoles [10–12], Scheme 7.1. The reaction involves a stepwise Cu(I)-catalyzed dipolar cycloaddition of a terminal acetylene to an organic azide. Azides ($-N_3$) and acetylenes ($-C\equiv CH$) are small, each just three atoms (C, H, or N), and are by definition kinetically stable, possessing high built-in energy, yet are metabolically stable. The click chemistry reaction between azides and acetylenes is biocompatible. It operates in water at ambient temperature, is tolerant to a broad range of pH values, and is bio-orthogonal—azides and acetylenes are inert in the biological milieu [13]. These reaction attributes have underpinned the recent remarkable application of click chemistry in bioimaging [14]. The favorable size and

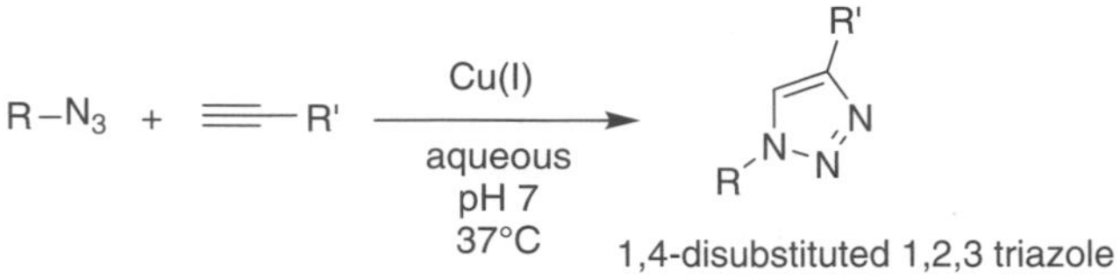

Scheme 7.1 Click chemistry synthesis of 1,4-disubstituted-1,2,3-triazoles by a 1,3-dipolar cycloaddition reaction of organic azides with terminal acetylenes.

inertness of azide and acetylene substrates have enabled their incorporation into biomolecules in living cells with minimal physiological perturbation, while subsequent chemoselective conjugation to small-molecule fluorescent probes allows the visualization and elucidation of highly specific cellular mechanisms [14]. Recent examples of click chemistry have reported a powerful advance that negates the need for copper catalysis, and click chemistry has been demonstrated within whole living cells without copper and its associated toxicity [15].

Both DCC and *in situ* click chemistry permit synthesis and screening to be combined into a single step with the target biomolecule guiding the assembly of fragments. A key differentiating attribute of these complementary concepts is that click chemistry utilizes an irreversible reaction to lock the fragments together, while DCC utilizes a reversible reaction to link fragments. Another important characteristic of *in situ* click chemistry is that the reaction avoids the combination of strong nucleophilic and electrophilic functional groups typical of DCC so that the reactive partner fragments for click chemistry are bio-orthogonal under virtually any reaction conditions [13]. In the absence of copper the click reaction occurs almost exclusively within the target's binding site, with minimal background reaction in the bulk solution, meaning that the formation of a triazole from fragments *in situ* almost guarantees that this triazole will be a potent lead compound [13]. A potential disadvantage of *in situ* click chemistry for drug lead discovery is the possibility that effective inhibitors are not assembled in the presence of the target and are thus "missed opportunities" in the screening campaign, that is, false negatives.

7.3. Dynamic Combinatorial Chemistry for Drug Lead Discovery

DCC has had considerable success with respect to the discovery of potent compounds for biomolecular targets. The entries in Table 7.1 list some of

Table 7.1   Published examples of DCC for drug lead discovery (2006 to Mid-2008)

Biomolecular target	Reversible chemistry	Maximum fragment size (Da)	Analytical method for hit identification
Carbonic anhydrase [17]	Acyl hydrazone exchange	<300	ESI-FTICR-MS
Metallo-β-lactamase [18]	Disulfide exchange	<200	ESI-MS
Transactivation-response element of HIV-1 [19]	Imine exchange	<230	SELEX, HPLC, and MALDI-TOF MS
α-1,3-Galactosyltrans-ferase; β-1,4-galactosyl-transferase [20]	Imine exchange	<250	HPLC
α-1,3-Galactosyltrans-ferase [21]	Imine exchange	<290	HPLC
Concanavalin A [22]	Disulfide exchange	~220	Quartz crystal microbalance
Galectin-3; *Viscum album* agglutinin; *Ulex europaeus* agglutinin [23]	Disulfide exchange	~220	Bioassay
Schistosoma japonica glutathione-S-transferase [24]	Conjugate addition of thiols to enones	~300	HPLC
Calmodulin [25]	Disulfide exchange	~300	HPLC
Subtilisin; albumin [26]	Disulfide exchange	~500	ESI-MS, HPLC
Aurora A kinase [27]	Disulfide exchange	<250	ESI-MS

the most recent published examples of DCC including 2006 to mid-2008. These publications exemplify the growing academic interest in the application of this informed chemistry for drug lead discovery. However, it is not apparent what may currently be occurring within the pharmaceutical industry, nor is it evident the progress that this approach has made toward advancing compounds into the drug discovery pipeline. The application of DCC for drug lead discovery within an industry setting was reported in 2002–2003 by Eliseev and coworkers, at then Therascope AG, leading to the discovery of very effective neuramindase inhibitors [16]. On the other

Table 7.2 Clinical and preclinical candidates derived from fragments

Biomolecular target	Candidate[a]	Company	Clinical progress
Fxa	LY-517717	Lilly/Protherics	Phase 2
PPAR agonist	PLX-204	Plexxikon	Phase 2
Bcl-X_L	ABT-263	Abbott	Phase 1/2a
Aurora	AT9283	Astex	Phase 1/2a
MMP-2 and -9	ABT-518	Abbott	Phase 1
B-Raf^{V600E}	PLX-4032	Plexxikon	Phase 1
MET	SGX523	SGX Pharmaceuticals	Phase 1
Aurora	SNS-314	Sunesis	Phase 1
CDKs	AT7519	Astex	Phase 1
HSP90	NVP-AUY922	Vernalis/Novartis	Phase 1
CDKs	AT9311/LCQ195	Astex/Novartis	Preclinical
PKB/Akt	AT13148	Astex	Preclinical
HSP90	AT13387	Astex	Preclinical
B-Raf^{V600E}	PLX-4720	Plexxikon	Preclinical
P38	RO6266	Roche	Preclinical
BCR-Abl^{T315I}	SGX393	SGX Pharmaceuticals	Preclinical

[a]Where disclosed compound structures are reported in Reference 4a.
Source: Reprinted from Reference 4a, with permission from American Chemical Society, Copyright (2008).

hand, a recent review on FBDD reports 16 preclinical Phase I and Phase II candidate compounds derived from fragments (Table 7.2) [4a]. It is clear that the progress level reported with fragments has not been forthcoming for DCC, and this begs the obvious question of "Why?". The success of translating novel fragments to candidates in the clinic may in part be attributed to the deliberate early-stage development of efficient screening techniques to identify fragments that bind only weakly to the target biomolecule [4]. Transforming DCC from a proof-of-concept academic investigation to an outcomes-focussed tool for use by the pharma industry similarly demands efficient screening methods to accurately identify the linked fragments of interest. This analytical hurdle is the focus of the remainder of this chapter.

7.4. Practical Considerations for Drug Discovery Applications of DCC

For the purpose of drug discovery we know that DCC can in principle circumvent the need to individually synthesize, characterize, and screen each possible library constituent. The chemistry, together with the target acting as a template, permits self-screening owing to selection and/or amplification of the "best binders" by molecular recognition events between fragments and the target. In the early days of DCC the integration of synthesis with screening was heralded as a key advantage of this approach [28]. What actually emerged as the standard DCL screening approach has deviated somewhat from this original notion, and instead an indirect method that screens for ligand enrichment has evolved. This indirect screening approach carries out the analysis of identical DCLs twice, that is, with and without the target biomolecule. The equilibrium concentration profiles of all library ligands when generated in both the presence and the absence of the target protein (usually following disruption of the ligand–target complexes) are compared, and the detection of ligand enrichment in the targeted DCL is the basis of identifying the "best binders". The complexity of screening with this indirect method increases with the size of the DCL owing to chromatographic (e.g., HPLC) and/or spectral (e.g., NMR) overlap so that either the need for synthesis of individual library components to validate assignments and/or the preparation and screening of DCL sublibraries becomes necessary for deconvolution. This is both labor- and time-intensive and has the effect of undermining the promoted advantages of DCC.

The synthetic advantages and target-informed product formation with DCC is of little consequence unless we have the ability to readily identify and characterize the linked fragments of interest. Accomplishing the efficient detection of these enriched products is clearly a make-or-break factor to DCC assuming a prominent place in the drug discovery pipeline for lead identification and/or optimization [29]. It would be a tremendous advantage if DCL screening methods could proceed without the need for chromatography, conversion to a static library, preparation of sublibraries, disruption of the protein–ligand complexes, synthesis of individual library ligands to validate analytical assignments or for the preparation and duplicate screening of identical DCLs (with and without the target). For DCC to have a significant impact on drug discovery campaigns necessitates direct screening protocols for rapid identification of ligands with affinity for the target protein, against the background of inactive DCL constituents.

Only recently has the original promise to integrate DCC synthesis and screening against a protein target been reported. These direct screening approaches have used either mass spectrometry [16,17,26] or X-ray crystallography [30] and represent a major advance toward addressing the difficult and impractical indirect analysis that has come to be associated with drug discovery applications of DCC. X-ray crystallography is without doubt unrivaled with regard to the resolution of structural information acquired; however, it is typically a slow turnaround experiment with a relatively high consumption of protein and so from a primary screening viewpoint somewhat restricted. By describing some recently published examples we will demonstrate that mass spectrometry allows direct screening of DCLs requiring only minimal protein consumption. Mass spectrometry is capable of generating a rapid result that yields the mass (and identity) of any DCL constituent that is noncovalently bound to the target protein. The theoretical product distribution within DCLs under a variety of library scenarios has been modeled, and this material has been reviewed elsewhere [31,32]. The models confirm the intuitive reduction in the observed concentration of the best binders that occurs with increasing number of DCL fragments [31]. This drop in concentration is not as sharp as one might have expected, so that amplification levels for the good binders are likely to remain within the detection capability of modern mass spectrometry equipment that is continuously improving in performance.

7.5. Mass Spectral Analysis of Proteins

Electrospray ionization mass spectrometry (ESI-MS) is generally used to accomplish the transfer of intact biological macromolecules such as proteins from solution into the gas phase. In ESI-MS, charged molecules in a solution are desolvated by a process that involves the formation of fine droplets generated by an applied high voltage [33]. The fine droplets successively decrease in size by evaporation of solvent, and as the charge density in the droplet becomes higher, the ejection of yet smaller droplets occurs. This process continues until a final stage of desolvation by evaporation leaves fully desolvated, charged molecules in the gas phase that are then guided into the mass analyzer of the mass spectrometer by a series of ion guides and/or transfer ion optics. The mass spectrometer then determines the mass-to-charge ratio (m/z) for the desolvated, charged molecules. Commonly used mass analyzers such as quadrupoles and ion traps have optimal performance in the m/z range up to 3000. As the parameter being measured is not the mass (m), but the mass-to-charge ratio (m/z), it is possible to analyze large molecular weight species such as proteins owing to multiple charging. For example, to bring the m/z value of a protein with a mass of 30 kDa into the optimal m/z range of a typical mass spectrometer, at least 10 charges must be incorporated on the protein in the gas phase. When using ESI-MS for molecular weight determination, samples are typically dissolved in a denaturing 1:1 water:organic solvent (acetonitrile or methanol) system containing ~0.1% acetic or formic acid. Under these conditions the protein unfolds to expose many of the amino acid side chains to the bulk solvent. The trace acid insures that many basic residues on the protein are protonated thus maximizing the signal obtained from the protein when running the mass spectrometer in positive ion mode. The mass spectrum for denatured carbonic anhydrase (CA) is shown in Fig. 7.1a. The spectrum is characterized by a broad envelope of charge states, from $+35$ to $+20$, corresponding to the family of differently charged molecular species present in solution. By contrast, in solution conditions capable of maintaining the native protein conformation (and specific noncovalent interactions thereof), many of the acidic and basic side chains are involved either in strong hydrogen-bonding interactions that help maintain the native-state protein structure or are otherwise inhibited from charging by their position in the protein structure (Fig. 7.1b). The fully folded protein typically shows a very narrow m/z distribution with just a few charge states observed and with a maximum in the distribution at a much higher m/z value than that for the denatured protein. This is shown for CA (Fig. 7.1b) where only three charge states are observed

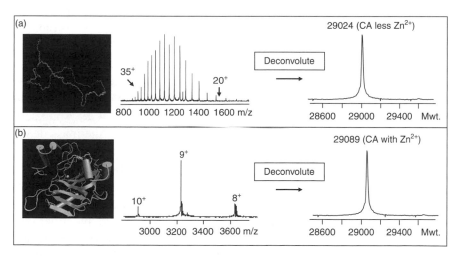

Figure 7.1   (**a**) The denatured conformation of the zinc metalloenzyme carbonic anhydrase and the ESI mass spectra obtained under acidic denaturing conditions. (**b**) The ESI mass spectra obtained under native-state conditions. The deconvoluted ESI mass spectra of carbonic anhydrase reveals the protein molecular weight. The three dimensional structure is protein Data Bank ID 1BN1.

($+10$, $+9$, and $+8$), compared with the more extensive charge state distribution under denaturing conditions. All modern ESI-equipped mass spectrometers include sophisticated deconvolution software to automatically and routinely convert MS data of protein samples to a zero charge state spectrum to generate the protein molecular weight; this is known as the deconvoluted mass spectrum. The deconvoluted mass spectrum will result in the same protein molecular weight from either denaturing or native state mass spectra (Fig. 7.1). Deconvolution calculations generally require only a fraction of a second for even quite complicated mass spectra.

7.6. ESI-MS for the Analysis of Noncovalent Protein–Ligand Interactions

ESI-MS has been used extensively to study proteins and complexes of proteins with naturally occurring substrates, inhibitors, and drugs. An extensive literature on this topic (more than 300 publications and numerous review articles as pointed out in a recent review [34]) indicates that while there are pitfalls, it is generally straightforward to adjust the parameters of the ESI-MS measurement so that the signals measured in the mass spectrometer for a protein, its ligands, and the noncovalent complexes thereof reflect the equilibrium concentrations of these species in solution. If an inhibitor is combined with its target protein, then the noncovalent complex of [protein + inhibitor] is observed in the ESI mass spectrum (usually together with some unbound protein) (Fig. 7.2). The mass difference between the peaks for the unbound protein and the protein–inhibitor complex ($\Delta m/z$) can be multiplied by the charge state to give directly the molecular weight of the binding inhibitor, that is, $MW_{inhibitor} = \Delta m/z \times z$ (Fig. 7.2). An exception to unique identification of inhibitor will occur only if the mass determined for the bound ligand can be attributed to isobaric ligands; these are structural isomers that share the same molecular formula and thus share an identical molecular mass and isotope distribution fingerprint. The contribution of these ambiguous constituents to a "hit" identified in the MS screen can be readily differentiated if required, for example, by either tandem MS experiments or "knockout" MS experiments. To confirm correct ESI-MS parameter adjustment, parameters for the ESI-MS measurement are generally first optimized on a target with a ligand of known binding constant such that the known binding constant is observed in the results [35]. Methods for simultaneously screening multiple binders require that the interaction between the target and any component in the screen be independent of the interactions with other components. This requirement

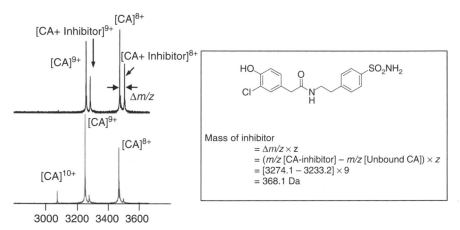

Figure 7.2 The mass-to-charge difference ($\Delta m/z$) between the peaks for the unbound carbonic anhydrase (CA) protein and the CA–inhibitor complex can be multiplied by the charge state to give directly the molecular weight of the CA-binding inhibitor, that is, $MW_{inhibitor} = \Delta m/z \times z$.

can be met by using a sufficient excess of the target in each screen [34]. Thus, when screening ligand libraries with unknown binding affinities, it is possible to include a small amount of compound with a known binding constant (K_d) to act as an internal calibrant for determining the K_d values for those library ligands for which noncovalent complexes with the target are observed [34]. Mass spectrometry has very high sensitivity and a wide dynamic range, and previous results have shown that by using these techniques it is possible to detect and determine K_d values for noncovalent complexes in the range 10 nM to 1 mM [34].

An important consideration for the success of direct observation of specific noncovalent complexes of native protein with small molecules using ESI-MS are the sample parameters: typically aqueous, near physiological pH, and with an ionic strength capable of maintaining the native target conformation and specific noncovalent interactions. Sample "contaminants" from an ESI-MS viewpoint include nonvolatile buffers, salts, and detergents as these sample components can suppress the ion abundance from the target of interest, leading to a poor mass spectrum result. Volatile solution components that do not form gas phase adducts with the target biomolecule are essential, and the most investigated and proven reliable volatile components for this purpose include the salts ammonium acetate and ammonium bicarbonate [36]. Protein samples are generally prepared in water with ammonium acetate or ammonium bicarbonate present in a concentration between 5 and 500 mM, depending on the protein. Inline techniques such as desalting, buffer exchange, and size exclusion chromatography have been routinely employed as a sample cleanup directly prior to ESI-MS analysis. This permits protein sample solutions to be first exchanged into ESI-compatible buffers to replace nonvolatile solution components while preserving protein–ligand noncovalent complexes.

A number of different types of ESI sources, known as nanospray sources, have been designed that can operate at lower sample flow rates ($10–200$ nL min^{-1}). These generate smaller droplets and improve the signal intensity of the protein–ligand noncovalent complexes further, with the added benefit of reducing protein consumption up to 100-fold compared to standard ESI flow rates. Nanospray has also been reported to be more tolerant to nonvolatile cations in solution [37]. Recently, an automated fabricated chip nanospray source has been developed. This chip-based device has improved the ease-of-use for nanospray, while the design eliminates carryover effects as the spray is produced directly from an orifice on each sample well of the chip [38].

7.7. Mass-Spectrometry-Based Screening of DCLs

7.7.1. Case Study 1: Carbonic Anhydrase

We first described the application of mass spectrometry to the direct screening of a DCL, wherein a protein was the target in 2006 [17]. This study applied electrospray ionization Fourier transform mass spectrometry (ESI-FTMS) to the screening of a DCL against the Zn(II) metalloenzyme carbonic anhydrase II (CA II). CA catalyzes the reversible hydration of carbon dioxide (CO_2) to give bicarbonate (HCO_3) and a proton (H^+). This is a fundamental physiological reaction that underpins essential cellular processes including respiration and transport of CO_2/HCO_3, provision of HCO_3 for biosynthesis, and electrolyte/fluid secretion [39]. Inhibition of CAs is implicated as a target for a range of disease states including cancer, glaucoma, and obesity [39]. The classical and minimal CA recognition fragment is an aromatic sulfonamide ($Ar–SO_2NH_2$), and the deprotonated sulfonamide group ($Ar–SO_2NH^-$) serves as a zinc-binding function in the active site of CAs. When this anchor fragment is further derivatized, then inhibitors with optimized target affinity, selectivity, and other pharmaceutical properties may be generated [39].

In this ESI-FTMS screening proof-of-concept experiment, a DCL was generated using the hydrazone exchange reaction from two hydrazide fragments (**1** and **2**) and five aldehyde fragments (**A–E**) (Scheme 7.2).

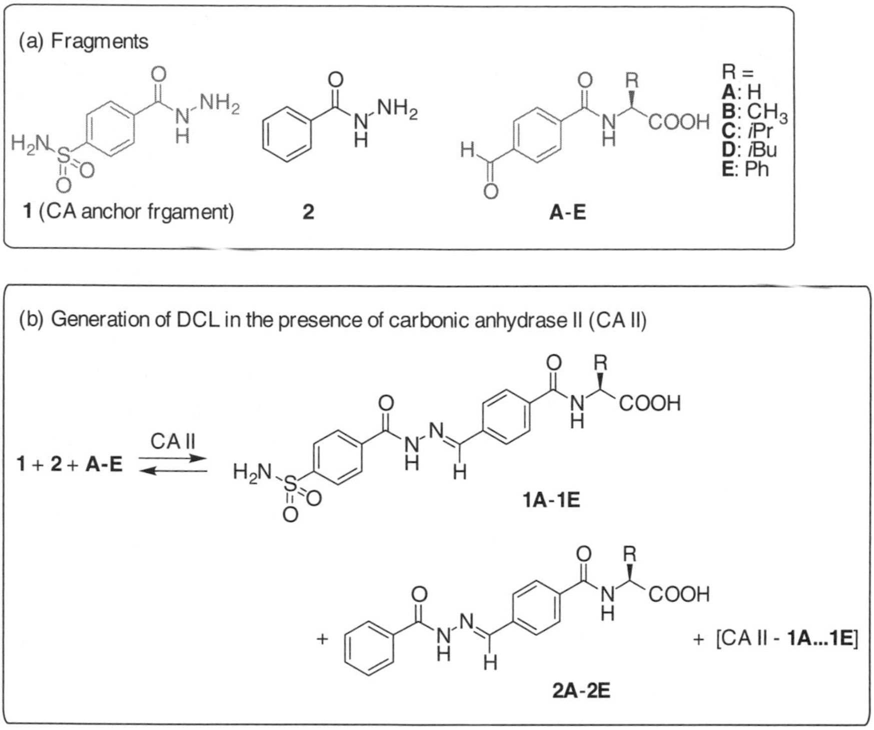

Scheme 7.2 Dynamic combinatorial chemistry targeting carbonic anhydrase. (a) Fragments for hydrazone exchange. (b) Dynamic combinatorial library generation.

Hydrazide **1** was designed as the CA anchor fragment and hence necessitated dual functionality: an $Ar–SO_2NH_2$ moiety for reliable CA affinity and a hydrazide moiety to take part in hydrazone exchange. Hydrazide **2** lacked the sulfonamide moiety of **1**, but was still able to participate in hydrazone exchange and thus functioned as a control compound. Fragments **A–E**, the exchange partners for **1** and **2**, were selected to introduce an array of "tails" onto **1** to enable exploration of periphery recognition interactions with CA.

A DCL of acyl hydrazones (**1A–1E**, **2A–2E**) was generated in the presence of CA II. The fragments and target were combined and incubated as follows: CA II (30 μM), **1** (15 μM), **2** (15 μM), and **A–E** (6 μM each); in 10 mM NH_4OAc (pH 7.2) with 1% DMSO to effect solubility, at 37°C for 40 hours (Scheme 7.2). The final DCL volume was 50 or 100 μL. The total accessible hydrazone concentration was equivalent to the concentration of CA II (30 μM), while individual hydrazone products could be formed to a maximum concentration of 6 μM or 20 mol% relative to CA II.

The mass spectrometry for this analysis was performed on an APEX® III 4.7 T FTICR mass spectrometer (Bruker Daltonics, Billerica, MA) fitted with an Apollo™ (Bruker Daltonics, Billerica, MA, USA) ESI source operated in negative ion mode. Broadband excitation was used to analyze a mass range from m/z 100 to 4500, with instrument parameters tuned and optimized for detecting m/z ~3000. DCL samples were infused directly into the ESI source at 2 µL min^{-1}. ESI-FTMS analysis of a solution containing only CA II (~29 kDa) yielded the ESI negative ion mass spectrum of Fig. 7.3a. Peaks corresponding to the -10 to -8 charge states of CA II were observed. This charge state envelope (low charge states and few charge states) is typical for CA II when in a compact, tightly folded native structure [35,40]. The mass spectrum of the CA II-DCL solution is presented in Fig. 7.3b. The same charge state envelope as for free CA II (Fig. 7.3a) was observed; however, each charge state now consisted of a grouping of peaks: a peak that corresponded to native CA II, and at higher m/z value a group of peaks that corresponded to the five different CA II–hydrazone noncovalent complexes (CA II–1A . . . 1E) as well as a small amount of (CA II–1). The amino functional groups on proteins are predominantly protonated in the pH range of this experiment, and potential imine products from the reaction of aldehyde fragments with amino groups on the protein target were not observed.

To further demonstrate the power of MS for integrating DCC synthesis and screening the MS/MS technique was next employed to verify the identity of the CA-bound ligands from the DCL. MS/MS experiments were performed using argon as the collision gas. The collision energy for the experiment was tuned to cause dissociation of the noncovalent protein–ligand complexes. The parent ions bearing the -9 charge state were selected and collision-activated dissociation yielded free CA II (both -8 and -9 charge states) and, important for the application to DCC screening, singly charged negative ions for the hydrazone ligands **1A–1E**, now well resolved by molecular mass (Fig. 7.3c). The masses of these ions were consistent with the $[M - H]^-$ ions expected for the DCL sulfonamide hydrazone products **1A–1E**. The sample quantity consumed for these MS/MS experiments was less than 100 µL, the initial ESI-FTMS experiment takes only minutes to perform, while the MS/MS experiment can be completed within 30 minutes. Confirmation of the results of this DCL experiment was then obtained by conducting a conventional solution-phase competitive binding assay for CA II to measure the equilibrium dissociation constants (K_i's) for the compounds described in this study. The DCL products **1A–1E** each exhibited increased affinity for the enzyme (K_i range 10.6–82.3 nM) when compared to the anchor fragment **1** ($K_i = 150$ nM).

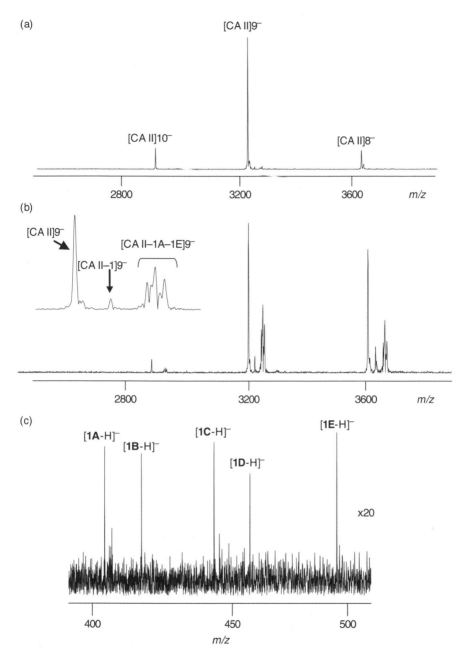

Figure 7.3 (a) ESI-FTMS negative ion mass spectrum of CA II (30 μM) in 10 mM NH₄OAc solution, 1% DMSO. (b) ESI-FTMS negative ion mass spectrum of a mixture of CA II (30 μM) and DCL containing 10 possible hydrazone products (**1A–1E** and **2A–2E**) in 10 mM NH₄OAc, 1% DMSO. (c) MS/MS spectrum obtained following collision-activated dissociation (CAD) of (CA II-hydrazone) noncovalent complexes to identify the hydrazones. This figure was published in Journal of the American Society for Mass Spectrometry, 17, Poulsen S.-A., Direct Screening of a Dynamic Combinatorial Library Using Mass Spectrometry, 1074–1080, Copyright Elsevier (2006).

Follow-up ESI-FTMS experiments from our laboratory have been effective with a 10-fold reduction of CA II concentration (from 30 to 3 μM, ca. ~4 μg protein/DCL based on 50 μL reaction volume) while retaining the ability to detect protein–ligand noncovalent complexes with good sensitivity down to 5 mol% ligand. Nanoelectrospray ionization (nanoESI), as described earlier in this chapter, has superior capabilities for investigating protein–ligand noncovalent complexes compared to standard ESI owing to lower sample flow rates (10–200 nL min^{-1}) and smaller droplets that improve the transfer of specific noncovalent complexes to the gas phase and hence increase signal intensity with the added benefit of reducing protein consumption. The application of nanospray FTMS to screening DCLs could allow screens to be carried out with subpicomole quantities of protein.

7.7.2. Case Study 2: Metallo-β-Lactamase

The application of mass spectrometry to the direct screening of a DCL has also been adopted more recently by others. Schofield and colleagues targeted the therapeutically relevant enzyme metallo-β-lactamase (BcII) from *Bacillus cereus* using DCC with disulfide exchange [18]. This enzyme (molecular weight ~25 kDa) catalyzes the hydrolysis of a range of clinically used β-lactam antibiotics and is of interest as a medicinal chemistry target owing to its involvement in the resistance of bacteria to antibiotics [41]. This DCC investigation stemmed from some earlier work by the same group in which it was confirmed, also using mass spectrometry, that BcII contains two active site zinc cations and that simple thiol fragments formed BcII–inhibitor complexes through a zinc-binding interaction [42].

In this study the fragment solutions were combined in an oxygen-free environment with the target enzyme BcII to give a DCL that consisted of BcII (15 μM), five anchor fragment dithiols (30 μM each), and 19 thiol "tail" fragments (10 μM each), in 15 mM NH$_4$OAc at pH 7.5. Each anchor fragment (**F–J**) had two thiol groups: one to facilitate Zn binding (thiol$_1$) and the second to participate in thiol–disulfide exchange for DCC (thiol$_2$). The DCC study examined 19 additional novel fragments: monothiols (**4–17** and **19–22**) and dithiol (**18**) as the coupling partners for **F–J** (Scheme 7.3).

Analysis was performed in a 96-well microtitre plate by ESI-MS on a Q-TOF mass spectrometer (Q-TOFmicro Micromass, Altrincham, UK) interfaced with a NanoMate™ chip-based nanoESI source (Advion Biosciences, Ithaca, NY). Samples were infused at a flow rate ~100 nL min^{-1}. Calibration and sample acquisition were performed in positive ion mode in the range of *m/z* 500–5000. The ESI-MS screen identified noncovalent complexes of BcII and confirmed the preference of native BcII for certain

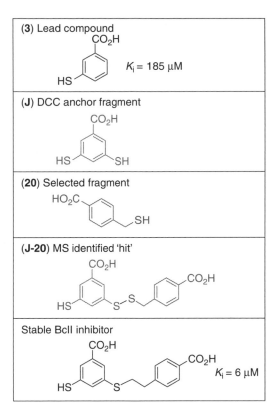

(3) Lead compound
CO_2H
$K_i = 185 \ \mu M$
HS

(J) DCC anchor fragment
CO_2H
HS   SH

(20) Selected fragment
HO_2C
SH

(J-20) MS identified 'hit'
CO_2H
CO_2H
HS   S—S

Stable BcII inhibitor
CO_2H
CO_2H
$K_i = 6 \ \mu M$
HS   S

(1) Carbonic anhydrase anchor fragment
O
N—NH_2
H
H_2N—S
O   O
$K_i = 150 \ nM$

(B) Target selected 'tail' fragment
O
N   COOH
H
H
O

(1B) MS identified 'hit'
O
O
N   COOH
H
H_2N—S   N—N
O   O   H   H
$K_i = 10.6 \ nM$

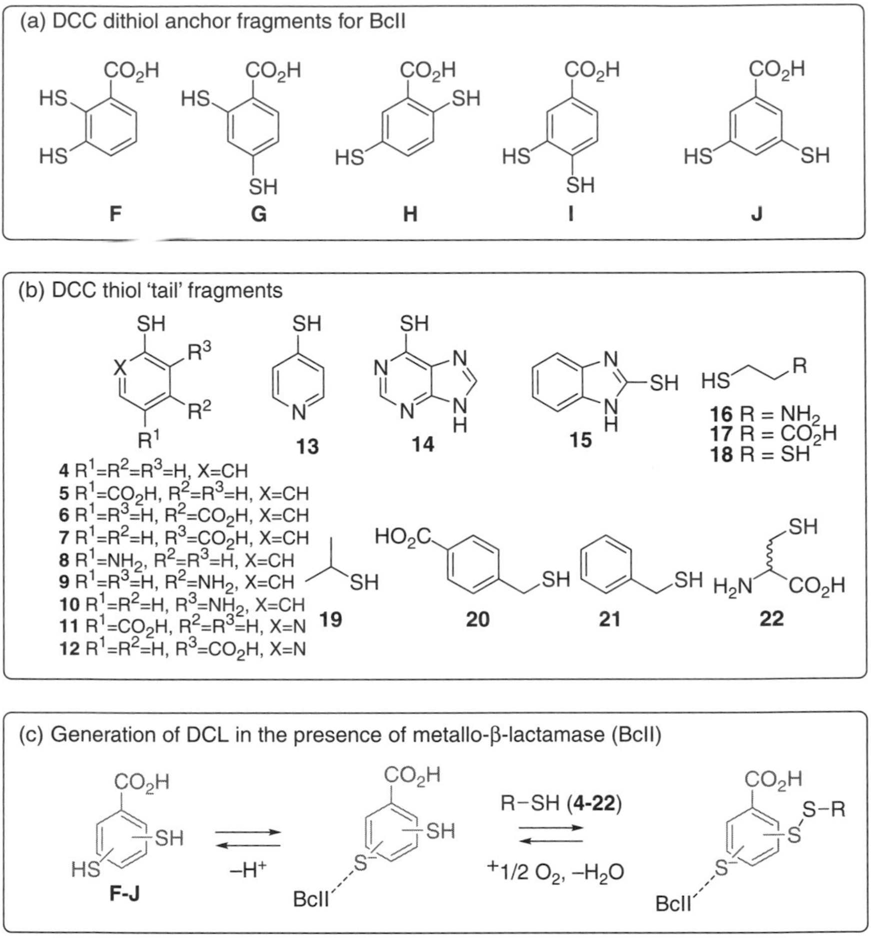

Scheme 7.3   (**a**) Anchor fragments equipped with two thiol moieties (**F–J**). (**b**) Thiol partner fragments (**4–22**). (**c**) DCC utilizing thiol-disulfide exchange and targeting metallo-β-lactamase (BcII).

disulfide heterodimers from all possible DCL constituents. The overall outcome for this study was the identification of four novel thiol compounds with K_i 6–17 μM, 10- to 30-fold more potent than the lead compound (**3**) with a K_i of 185 μM.

7.7.3. Case Study 3: Aurora A Kinase

Very recently Erlanson and colleagues described the discovery of aurora A kinase inhibitors using a purely dynamic chemistry approach [27] that

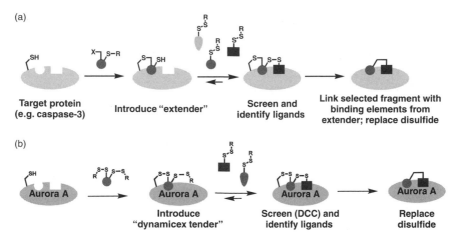

Figure 7.4 (**a**) Schematic illustration of the tethering with covalent extenders technique. Reprinted from Reference 43, with permission from Macmillan Publishers Ltd, Copyright (2003). (**b**) Schematic illustration of the site-specific DCC technique. Reprinted from Reference 27, with permission from Elsevier, Copyright (2008).

complements and advances earlier work on "Tethering with covalent extenders" by this group [43] (Fig. 7.4a). Kinases have in common a purine binding site and neighboring nonconserved adaptive region. Exploiting binding to the adaptive region allows highly specific kinase inhibitors to be developed. The recent work targeting aurora A kinase differs from the group's earlier work, as the *irreversible* covalent capture of the extender fragment by the target (Fig. 7.4a) is now replaced with a *reversible* capture strategy using the disulfide exchange reaction (Fig. 7.4b). Dynamic chemistry thus plays a dual role in the approach: first to capture and anchor the extender fragment within the target; and second to build upon the extender fragment structure. Figure 7.4b illustrates a schematic of this site-specific DCC approach.

A construct of aurora A kinase was prepared in which a cysteine residue was incorporated near the ATP-binding site. A dynamic extender with a diaminopyrimidine core and two disulfide-containing appendages was synthesized (**23**). The pyrimidine core recognizes the ATP-binding site and so may anchor to aurora A kinase through reaction of one disulfide-containing arm with the introduced cysteine residue of this target. The other extender arm is then positioned to capture disulfide-containing fragments that possess affinity for the adaptive binding site. A total of ~4500 disulfide-containing fragments were then screened, with ESI-MS used as the primary screening tool similar to that described above, in order to quickly identify the protein–ligand complexes formed and stabilized by aurora A kinase. The extender (**23**), aurora A kinase, and disulfide fragments were combined in a 96-well plate format as follows: aurora A kinase (5 μM, 50 mM Tris, pH 8); 1 mM 2-mercaptoethanol; dynamic extender (50 μM); and 10 unique disulfide-containing fragments (50 μM each). After 4 hours at room temperature, each well was subjected to LC-MS analysis using an LCT time-of-flight mass spectrometer equipped with an eight-channel parallel multiplexed (MUX) ESI interface (Waters Corp., Milford, MA) and a Gilson 215/889 eight-channel liquid handler (Gilson, Middleton, WI). The protein samples were desalted prior to MS analysis using reverse-phase Protein lTraps (Michrom BioResources Inc., Auburn, CA). Protein charge state distributions were deconvoluted to obtain the zero-charge spectrum. Less than five 96-well plates were required to complete the analysis of ~4500 fragments. An example of the screening output is shown in Fig. 7.5; this mass spectrum shows a peak at mass 32,012 Da that corresponds to aurora A kinase linked to extender **23** and fragment **24**. This hit (**23** + **24** → **25**) was followed up by the direct linking of the extender and fragment (**26**) followed by conventional optimization strategies resulting in a new lead compound (**27**) with an IC_{50} of 2.9 μM. The binding mode for compound **27** was then verified to target both the purine-binding site and the adaptive region of the kinase target using protein X-ray crystallography [27].

7.8. Summary and Outlook

Proteins exist as an ensemble of equilibrating conformers, and variations in structure can range from subtle to extreme [44] and invariably compromise the outcome of structure-based lead optimization efforts that are typically guided by a static and/or ensemble-averaged conformation. The principles of DCC in the context of target-guided synthesis are testament to both conceptual and synthetic advance that, while currently a niche component of medicinal chemistry, has the potential to impact on the direction of modern drug discovery campaigns. The reported DCC medicinal chemistry successes have generally stemmed from a prior knowledge of a reliable target recognition fragment or "anchoring" fragment. This anchor fragment is then furnished with the necessary functional group(s) to participate in DCC, typically with a much larger panel of complementary functionalized fragments. The *in situ* chemistry is free to scan the active site architecture of the target to link those fragments that can best exploit molecular recognition interactions with the target. As DCC continues to

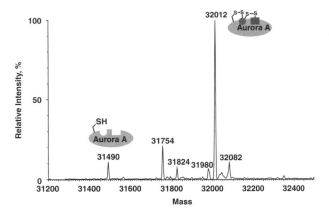

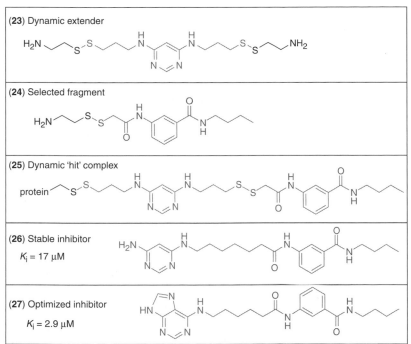

Figure 7.5 Deconvoluted, zero charge state mass spectrum demonstrating a hit from a DCL-targeting metallo-β-lactamase (BcII). The dominant peak corresponds to aurora A kinase linked to extender **23**, which is in turn linked to fragment **24** to give **25** (dynamic "hit"). Reprinted from Reference 27, with permission from Elsevier, Copyright (2008).

develop, this field should aim to deliver results driven entirely from novel anchor fragments identified by independent methods.

The immediate challenge that presents is to demonstrate practicality of DCC in a drug discovery setting and to engage as appropriate with big pharma. The selection and assembly of fragments in a DCL can in principle be screened by an activity assay or by any number of analytical techniques; however, for drug discovery it is a combination of speed, miniaturization, and ultimately the nature of the information sought that will drive the choice of analysis. The candidates in clinical trials that have stemmed from FBDD, while overshadowing the limited success so far for DCC, may also inspire the future direction of this field. Fragments discovered through fragment-based screening are clearly well poised to lead the DCC campaigns for target-guided fragment optimization, and it is proposed that the currently small overlapping footprint of FBDD with DCC will grow leading to a powerful and perhaps core future methodology. Stepping outside our comfort zone and shedding the "need to know everything" philosophy of academic chemistry and in its place settle for what we "really need to know" for medicinal chemistry (i.e., the molecular structure of the linked fragments that are selected and amplified in the DCC system under investigation) may represent an appropriate way to advance towards the goal of drug discovery.

In general, the speed, simplicity, and sensitivity of mass spectrometry based screening of ligands against biomolecular targets makes it an excellent choice for a primary screen. This is especially true for DCC applications wherein the identity of the binding species is not initially known, except that any binding species must be composed of the building block fragments employed in the experiment. The mass spectral detection of protein–ligand complexes readily allows determination of the mass of the binding ligand, which can be used to ascertain the building block fragments comprising the ligand. On this basis the integration of DCC with a mass spectrometry based screen should fulfil the screening requirements and provide a very effective means for identifying the combination of fragments that bind to a given target under biologically relevant conditions. Ligands identified by mass spectrometric screening may then be verified by more traditional biological activity secondary screening tests. As nanoESI sources become more user friendly it is expected that more DCL applications will emerge that utilize MS for screening biomolecular targets to identify binding partners. In addition coupling nanoESI with automated sample handling could significantly reduce both the time and target quantity required to facilitate analysis, this is clearly desirable for drug discovery applications. When using MS techniques it is reasonable to expect that increased DCL size need not increase the complexity of this screening protocol, owing to the sensitivity, high resolution and MS/MS capabilities which should avoid the need for multiple sub-libraries for deconvolution of larger DCLs.

The analytical demands of compound libraries has directed the attention of the pharmaceutical industry towards mass spectrometry owing to attributes of speed, intrinsic sensitivity, specificity, low sample consumption and the capability of resolving vast numbers of compounds in complex compound mixtures. The examples in this chapter demonstrate that there is enormous scope for ESI mass spectrometry as a direct and primary screening tool for the identification of small molecules formed by DCC in the presence of a protein target. The ESI mass spectrometry screening has permitted concurrent identification of all ligands of interest through direct analysis. The approach distinguished the effective from ineffective (which we do not necessarily need to know about) combination of building blocks in the DCL by specific detection of the target protein–ligand non-covalent complexes and in all examples novel ligands were identified with improved enzyme inhibition properties compared to the lead fragments.

Abbreviations

1,3-DCR: 1,3-dipolar cycloaddition reaction
BcII: metallo-β-lactamase
CA: carbonic anhydrase
DCC: dynamic combinatorial chemistry
DCL: dynamic combinatorial library
ESI: electrospray ionization
FBDD: fragment-based drug discovery
FTICR: Fourier transform ion cyclotron resonance
K_i: inhibitor equilibrium dissociation constant
K_d: ligand equilibrium dissociation constant
MS: mass spectrometry

Acknowledgment

A portion of this chapter has appeared in a similar format in a previous publication [45]. The authors thank Dan Erlanson for helpful discussion on the material presented herein.

References

1. (a) Frantz, S. Pipeline problems are increasing the urge to merge. *Nat. Rev. Drug Discov.* **2006**, *5*, 977–979. (b) Houlton, S. Breaking the rules. *Chem. World* **2008**, *5*, 58–62.

2. (a) Hughes, B. Pharma pursues novel models for academic collaboration. *Nat. Rev. Drug Discov.* **2008**, *7*, 631–632. (b) Gray, N. S. Drug discovery through industry–academic partnerships. *Nat. Chem. Biol.* **2006**, *2*, 649–653. (c) Nwaka, S.; Hudson, A. Innovative lead discovery strategies for tropical diseases. *Nat. Rev. Drug Discov.* **2006**, *5*, 941–955.

3. (a) Hopkins, A. L.; Groom, C. R. *Nat. Rev. Drug Discov.* **2002**, *1*, 727–730. (b) Lipinski, C. A.; Lombardo, F.; Dominy, B. W.; Feeney, P. J. *Adv. Drug Deliv. Rev.* **2001**, *46*, 3–26.

4. FBDD reviews: (a) Congreve, M.; Chessari, G.; Tisi, D.; Woodhead, A. J. Recent developments in fragment-based drug discovery. *J. Med. Chem.* **2008**, *51*, 3661–3680. (b) Erlanson, D. A. Fragment-based lead discovery: A chemical update. *Curr. Opin. Biotechnol.* **2006**, *17*, 643–652.

5. Huc, I.; Lehn, J.-M. Virtual combinatorial libraries: Dynamic generation of molecular and supramolecular diversity by self-assembly. *Proc. Natl. Acad. Sci. U.S.A.* **1997**, *94*, 2106–2110.

6. Lewis, W. G.; Green, L. G.; Grynszpan, F.; Radic, Z.; Carlier, P. R.; Taylor, P.; Finn, M. G.; Sharpless, K. B. Click chemistry *in situ*: Acetylcholinesterase as a reaction vessel for the selective assembly of a femtomolar inhibitor from an array of building blocks. *Angew. Chem. Int. Ed.* **2002**, *41*, 1053–1057.

7. Corbett, P. T.; Leclaire, J.; Vial, L.; West, K. R.; Wietor, J.-L.; /Sanders, J. K. M.; Otto, S. Dynamic combinatorial chemistry. *Chem. Rev.* **2006**, *106*, 3652–3711.

8. Cousins, G. R. L.; Poulsen, S.-A.; Sanders, J.K.M. Molecular evolution: Dynamic combinatorial libraries, autocatalytic networks and the quest for molecular function. *Curr. Opin. Chem. Biol.* **2000**, *4*, 270–279.

9. Ramström, O.; Lehn, J.-M. Dynamic ligand assembly. In: *Comprehensive medicinal chemistry II (Volume 3)*, Taylor, J.B.; Triggle, D.J. editors. Elsevier Ltd., Oxford, UK, pp. 959–976, **2006**.

10. Huisgen, R.; Szeimies, G.; Moebius, L. 1,3-Dipolar cycloadditions XXXII. Kinetics of the addition of organic azides to carbon–carbon multiple bonds. *Chem. Ber.* **1967**, *100*, 2494–2507.

11. Tornøe, C. W.; Christensen, C.; Meldal, M. Peptidotriazoles on solid phase: [1,2,3]-Triazoles by regiospecific copper(I)-catalyzed 1,3-dipolar cycloadditions of terminal alkynes to azides. *J. Org. Chem.* **2002**, *67*, 3057–3064.

12. Rostovtsev, V. V.; Green, L. G.; Fokin, V. V.; Sharpless, K. B. A Stepwise Huisgen Cycloaddition Process: Copper(I)-Catalyzed Regioselective "Ligation" of Azides and Terminal Alkynes. *Angew. Chem. Int. Ed.* **2002**, *41*, 2596–2599.

13. Sharpless, K. B.; Manetsch, R. *In situ* click chemistry: A powerful means for lead discovery. *Expert Opin. Drug Discov.* **2006**, *1*, 525–538.

14. (a) Sawa, M.; Hsu, T.-L.; Itoh, T.; Sugiyama, M.; Hanson, S. R.; Vogt, P. K.; Wong, C.-H. Glycoproteomic probes for fluorescent imaging of fucosylated glycans in vivo. *Proc. Natl. Acad. Sci. U.S.A.* **2006**, *103*, 12371–12376. (b) Rabuka, D.; Hubbard, S. C.; Laughlin, S. T.; Argade, S. P.; Bertozzi, C. R. Chemical reporter strategy to probe glycoprotein fucosylation. *J. Am. Chem. Soc.* **2006**, *128*, 12078–12079. (c) Baskin, J. M.; Prescher, J. A.; Laughlin, S. T.; Agard, N. J.; Chang, P. V.; Miller, I. A.; Lo, A.; Codelli, J. A.; Bertozzi, C. R. Copper-free click chemistry for dynamic in vivo imaging. *Proc. Natl. Acad. Sci. U.S.A.* **2007**, *104*, 16793–16797. (d) Gupta, S. S.; Kuzelka, J.; Singh, P.; Lewis, W. G.; Manchester, M.; Finn, M. G. Accelerated bioorthogonal conjugation: A practical method for the ligation of diverse functional molecules to a polyvalent virus scaffold. *Bioconjug. Chem.* **2005**, *16*, 1572–1579.

15. (a) Laughlin, S. T.; Baskin, J. M.; Amacher, S. L.; Bertozzi, C. R. In vivo imaging of membrane-associated glycans in developing zebrafish. *Science* **2008**, *320*, 664–667. (b) Ning, X.; Guo, J.; Wolfert, M. A.; Boons, G.-J. Visualizing metabolically labeled glycoconjugates of living cells by copper-free and fast Huisgen cycloadditions. *Angew. Chem. Int. Ed.* **2008**, *47*, 22553–22559.

16. a) Hochguertel, M.; Biesinger, R.; Kroth, H.; Piecha, D.; Hofmann, M. W.; Krause, S.; Schaaf, O.; Nicolau, C.; Eliseev, A. V. Ketones as building blocks for dynamic combinatorial libraries: Highly active neuraminidase inhibitors generated via selection pressure of the biological target. *J. Med. Chem.* **2003**, *46*, 356–358. (b) Hochguertel, M.; Kroth, H.; Piecha, D.; Hofmann, M. W.; Nicolau, C.; Krause, S.; Schaaf, O.; Sonnenmoser, G.; Eliseev, A. V. Target-induced formation of neuraminidase inhibitors from in vitro virtual combinatorial libraries. *Proc. Natl. Acad. Sci. U.S.A.* **2002**, *99*, 3382–3387.

17. Poulsen, S.-A. Direct screening of a dynamic combinatorial library using mass spectrometry. *J. Am. Soc. Mass Spectrom.* **2006**, *17*, 1074–1080.

18. Lienard, B. M. R.; Selevsek, N.; Oldham, N. J.; Schofield, C. J. Combined mass spectrometry and dynamic chemistry approach to identify metalloenzyme inhibitors. *ChemMedChem* **2007**, *2*, 175–179.

19. Bugaut, A.; Toulme, J.-J.; Rayner, B. SELEX and dynamic combinatorial chemistry interplay for the selection of conjugated RNA aptamers. *Org. Biomol. Chem.* **2006**, *4*, 4082–4088.

20. Valade, A.; Urban, D.; Beau, J.-M. Two galactosyltransferases' selection of different binders from the same uridine-based dynamic combinatorial library. *J. Comb. Chem.* **2007**, *9*, 1–4.

21. Valade, A.; Urban, D.; Beau, J.-M. Target-assisted selection of galactosyltransferase binders from dynamic combinatorial libraries: An unexpected solution with restricted amounts of enzymes. *Chembiochem* **2006**, *7*, 1023–1027.

22. Pei, Z.; Larsson, R.; Aastrup, T.; Anderson, H.; Lehn, J.-M.; Ramström, O. Quartz crystal microbalance bioaffinity sensor for rapid identification of

glycosyldisulfide lectin inhibitors from a dynamic combinatorial library. *Biosens. Bioelectron.* **2006**, *22*, 42–48.

23. Andre, S.; Pei, Z.; Siebert, H.-C.; Ramström, O.; Gabius, H.-J. Glycosyl-disulfides from dynamic combinatorial libraries as O-glycoside mimetics for plant and endogenous lectins: Their reactivities in solid-phase and cell assays and conformational analysis by molecular dynamics simulations. *Bioorg. Med. Chem.* **2006**, *14*, 6314–6326.

24. Shi, B.; Stevenson, R.; Campopiano, D. J.; Greaney, M. F. Discovery of glutathione S-transferase inhibitors using dynamic combinatorial chemistry. *J. Am. Chem. Soc.* **2006**, *128*, 8459–8467.

25. Milanesi, L.; Hunter, C. A.; Sedelnikova, S. E.; Waltho, J. P. Amplification of bifunctional ligands for calmodulin from a dynamic combinatorial library. *Chem. Eur. J.* **2006**, *12*, 1081–1087.

26. Danieli, B.; Giardini, A.; Lesma, G.; Passarella, D.; Peretto, B.; Sacchetti, A.; Silvani, A.; Pratesi, G.; Zunino, F. Thiocolchicine–podophyllotoxin conjugates: Dynamic libraries based on disulfide exchange reaction. *J. Org. Chem.* **2006**, *71*, 2848–2853.

27. Cancilla, M. T.; He, M. M.; Viswanathan, N.; Simmons, R. L.; Taylor, M.; Fung, A. D.; Cao, K.; Erlanson, D. A. Discovery of an aurora kinase inhibitor through site-specific dynamic combinatorial chemistry. *Bioorg. Med. Chem. Lett.* **2008**, *18*, 3978–3981.

28. Ganesan, A. Strategies for the dynamic integration of combinatorial synthesis and screening. *Angew. Chem. Int. Ed.* **1998**, *37*, 2828–2831.

29. Weber, L. *In Vitro* combinatorial chemistry to create drug candidates. *Drug Discov. Today Technol.* **2004**, *1*, 261–267.

30. Congreve, M. S.; Davis, D. J.; Devine, L.; Granata, C.; O'Reilly, M.; Wyatt, P. G.; Jhoti, H. Detection of ligands from a dynamic combinatorial library by X-ray crystallography. *Angew. Chem. Int. Ed.* **2003**, *42*, 4479–4482.

31. Corbett, P. T.; Otto, S.; Sanders, J. K. M. What are the limits to the size of effective dynamic combinatorial libraries? *Org. Lett.* **2004**, *6*, 1825–1827.

32. Moore, J. S.; Zimmerman, N. W. "Masterpiece" copolymer sequences by targeted equilibrium-shifting. *Org. Lett.* **2000**, *2*, 915–918.

33. Fenn, J. B.; Mann, M.; Meng, C. K.; Wong, S. F.; Whitehouse, C. M. Electrospray ionization for mass spectrometry of large biomolecules. *Science* **1989**, *246*, 64–71.

34. Hofstadler, S. A.; Sannes-Lowery, K. A. Applications of ESI-MS in drug discovery: Interrogation of noncovalent complexes. *Nat. Rev. Drug Discov.* **2006**, *5*, 585–595.

35. (a) Cheng, X.; Chen, R.; Bruce, J. E.; Schwartz, B. L.; Anderson, G. A.; Hofstadler, S. A.; Gale, D. C.; Smith, R. D.; Gao, J.; Sigal, G. B.; Mammen, M.; Whitesides, G. M. Using electrospray ionization FTICR mass spectrometry to study competitive binding of inhibitors to carbonic anhydrase. *J. Am.*

Chem. Soc. **1995**, *117*, 8859–8860 (b) Gao, J.; Cheng, X.; Chen, R.; Sigal, G. B.; Bruce, J. E.; Schwartz, B. L.; Hofstadler, S. A.; Anderson, G. A.; Smith, R. D.; Whitesides, G. M. Screening derivatized peptide libraries for tight binding inhibitors to carbonic anhidrase II by electrospray ionization-mass spectrometry. *J. Med. Chem.* **1996**, *39*, 1949–1955 (c) Gao, J.; Wu, Q.; Carbeck, J.; Lei, Q. P.; Smith, R. D.; Whitesides, G. M. Probing the energetics of dissociation of carbonic anhidrase–ligand complexes in the gas phase. *Biophys. J.* **1999**, *76*, 3253–3260.

36. Loo, J. A. Electrospray ionization mass spectrometry: A technology for studying noncovalent macromolecular complexes. *Int. J. Mass Spectrom.* **2000**, *200*, 175–186.

37. Benkestock, K.; Sundqvist, G.; Edlund, P. O.; Roeraade, J. Influence of droplet size, capillary-cone distance and selected instrumental parameters for the analysis of noncovalent protein–ligand complexes by nanoelectrospray ionization mass spectrometry. *J. Mass Spectrom.* **2004**, *39*, 1059–1067.

38. (a) Keetch, C. A.; Hernánndez, H.; Sterling, A.; Baumert, M.; Allen, M. H.; Robinson, C. V. Use of a microchip device coupled with mass spectrometry for ligand screening of a multi-protein target. *Anal. Chem.* **2003**, *75*, 4937–4941 (b) Zhang, S.; Van Pelt, C. K.; Wilson, W. D. Quantitative determination of noncovalent binding interactions using automated nanoelectrospray mass spectrometry. *Anal. Chem.* **2003**, *75*, 3010–3018 (c) Benkestock, K.; Van Pelt, C. K.; Akerud, T.; Sterling, A.; Edlund, P. O.; Roeraade, J. Automated nanoelectrospray mass spectrometry for protein–ligand screening by noncovalent interaction applied to human H-FABP and A-FABP. *J. Biomol. Screen.* **2003**, *8*, 247–256.

39. (a) Supuran, C. T. Carbonic anhydrases: Novel therapeutic applications for inhibitors and activators. *Nat. Rev. Drug Discov.* **2008**, *7*, 168–181. (b) Supuran, C. T. Carbonic anhydrases as drug targets: An overview. *Curr. Top. Med. Chem.* **2007**, *7*, 825–833. (c) Supuran, C. T.; Scozzafava, A. Carbonic anhydrases as targets for medicinal chemistry. *Bioorg. Med. Chem.* **2007**, *15*, 4336–4350. (d) Winum, J.-Y.; Poulsen, S.-A.; Supuran, C. T. Therapeutic applications of glycosidic carbonic anhydrase inhibitors. *Med. Res. Rev.* **2009**, *29*, 419–435.

40. Smith, R. D.; Bruce, J. E.; Wu, Q.; Lei, P. New mass spectrometric methods for the study of noncovalent associations of biopolymers. *Chem. Soc. Rev.* **1997**, *26*, 191–202 (and references therein).

41. Bush, K.; Jacoby, G. A.; Medeiros, A. A. A functional classification scheme for β-lactamases and its correlation with molecular structure. *Antimicrob. Agents Chemother.* **1995**, *39*, 1211–33.

42. Selevsek, N.; Tholey, A.; Heinzle, E.; Lienard, B. M. R.; Oldham, N. J.; Schofield, C. J.; Heinz, U.; Adolph, H.-W.; Frere, J.-M. Studies on ternary metallo-β-lactamase inhibitor complexes using electrospray ionization mass spectrometry. *J. Am. Soc. Mass Spectrom.* **2006**, *17*, 1000–10004.

43. Erlanson, D. A.; Lam, J. W.; Wiesmann, C.; Luong, T. N.; Simmons, R. L.; DeLano, W. L.; Choong, I. C.; Burdett, M. T.; Flanagan, W. M.; Lee, D.; Gordon, E. M.; O'Brien, T. *In situ* assembly of enzyme inhibitors using extended tethering. *Nat. Biotechnol.* **2003**, *21*, 308–314.

44. Min, W.; English, B. P.; Luo, G.; Binny, J. Cherayil, B. J.; Kou, S. C.; Xie, X. S. Fluctuating enzymes: Lessons from single-molecule studies. *Acc. Chem. Res.* **2005**, *38*, 923–931.

45. Poulsen, S.-A.; Kruppa, G. H. *In situ* fragment-based medicinal chemistry: Screening by mass spectrometry. In: *Fragment-based drug discovery: Practical aspects*, Zartler, E. R., Shapiro, M. J. editors. Wiley, London, Chapter 7, **2008**.

Chapter 8

Dynamic Combinatorial Methods in Materials Science

Takeshi Maeda, Hideyuki Otsuka, and Atsushi Takahara

8.1. Introduction

Recent advances in polymer chemistry have allowed almost absolute syntheses of polymers with desired molecular weights and narrow molecular weight distributions as well as syntheses of many kinds of polymers with a variety of functionalities [1]. Efforts for developments of novel polymers prepared by chain-growth polymerization [2–5] and step-growth polymerization [6] have focused on the precise production of uniform macromolecules. In contrast, a novel methodology for the synthesis of dynamic polymeric materials by means of "reversible" polymerization has recently been suggested for developing future materials related to their reversibility [7–10]. This method differs from conventional polycondensations, which are implemented by reversible reactions such as esterification and etherification, because they involve pursuing structural unity and avoid structural diversity. This set of schemes for the generation of dynamic polymeric materials is closely related to dynamic combinatorial chemistry (DCC). The dynamic polymeric material focuses on structural diversity in the process of the equilibrium reaction and on the changeable and tunable properties of the resulting polymer after polymerizations by external stimuli. In this methodology, polymerizations progress through dynamic combinatorial libraries (DCLs) [11–15] involving

Dynamic Combinatorial Chemistry, edited by Benjamin L. Miller
Copyright © 2010 John Wiley & Sons, Inc.

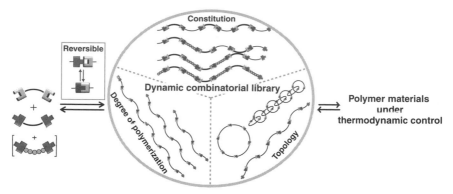

Figure 8.1   A dynamic combinatorial library on the preparation of polymer materials bearing reversible linkages.

multiple members having various molecular lengths and diversities of chemical compositions and topologies (Fig. 8.1). These library members are under equilibrium and seeking the most stable structures in the dynamic library. As a consequence, most stable polymers under thermodynamic control are spontaneously generated as major products.

DCC depends on a reversible connection process under mild conditions and an adequate time scale for the spontaneous production of library members consisting of basic components. Recently, the dynamic covalent bond having both reversibility and stability against the external environment has attracted much attention, as it can possibly be used in the development of novel materials in which the components can be reconstructed, deconstructed, or reshuffled, triggered by external stimuli [16]. Therefore, the dynamic covalent bond is advantageous for establishing a DCL, and for developing novel materials with a reversible nature.

This chapter will outline the synthesis of polymeric materials pursuing structural diversity and prepared by equilibrium reactions through DCLs. In particular, the dynamic covalent polymers will be focused upon because of their high stability and processability. In addition, advanced approaches to polymeric materials in DCC will be outlined. In this chapter, the authors will only discuss covalent polymers, excluding noncovalent polymers (supramolecular polymers) that can be found in References 7 and 8.

8.2. Dynamic Polymer Materials Prepared by Equilibrium Reactions

Dynamic polymers are defined as polymeric materials that can revert to other polymeric systems, cyclic or acyclic oligomers, or monomers under the thermodynamic control triggered by the external stimuli. Unlike conventional polymer materials, these can change their components, topologies, and molecular lengths after polymerization. These changeable and tunable polymers will be classified by the kinds of their dynamic covalent bonds.

8.2.1. C=N Exchange Reaction

C=N bonds such as imines and hydrazones, produced by amino/carbonyl condensations, are attractive in view of the wide range of structural variations available, the easy synthetic accessibility, and the control through conditions of yields, rates, and reversibility. Zhao and Moore have designed a supramolecular-assisted dynamic covalent polymer based on imine metathesis (Scheme 8.1) [17,18]. Curved tetrameric oligomers **1** and **2** bearing bis-imino end groups were synthesized in acetonitrile and chloroform in the presence of acid catalyst. These two solvents were used because the curved *m*-phenylene ethynylene oligomers tend to adopt an ordered helical conformation in polar acetonitrile and a random conformation state in apolar chloroform by the agency of solvophobic interactions [19]. Since these reactions were carried out in closed systems, a variety of oligomers with low molecular weights were expected to be generated with a monoimine by-product in the absence of a driving force to produce high molecular weight polymers. However, a high molecular weight polymer was generated in acetonitrile after achieving equilibrium, although low molecular weight oligomers were yielded in chloroform, likely due to the formation of helices. The folding energy for helix formation was the driving force for shifting the equilibrium to generate high molecular weight polymers. The resulting high molecular weight polymer in acetonitrile was allowed to reform either by the dilution of acetonitrile solution with chloroform or by changing the temperature. This sophisticated system depends on a reversible reaction in a closed system and helix formation as the driving force (Fig. 8.2).

Dynamic polymeric systems utilizing the C=N exchange reaction have been reported by Lehn's group. They have suggested a polymerization system consisting of a fluorene-based dialdehyde monomer **4**, cyclohexane diamine **5**, and fluorene-based diamine **6** as a comonomer (Scheme 8.2) [20,21]. In principle, a 1:1:1 mixture of all monomers in ethanol was expected to yield the two-component polymers **7** and **8** together with all component-mixed polymers. However, polymer **7** was dominantly yielded (80%) due to the nucleophilicity of diamines. The nucleophilicity of aliphatic diaminocyclohexane is much higher than that of aromatic

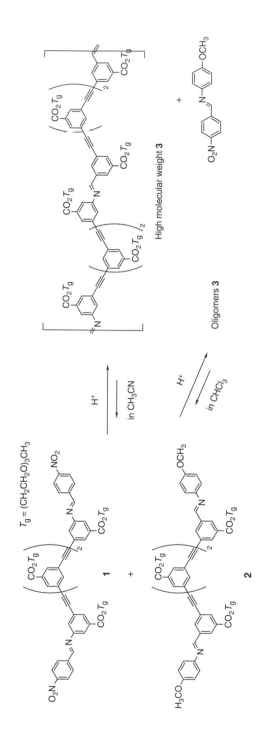

Scheme 8.1 Imine metathesis polymerizations of **1** and **2** in CH$_3$CN and CHCl$_2$, and their reversible feature [17,18].

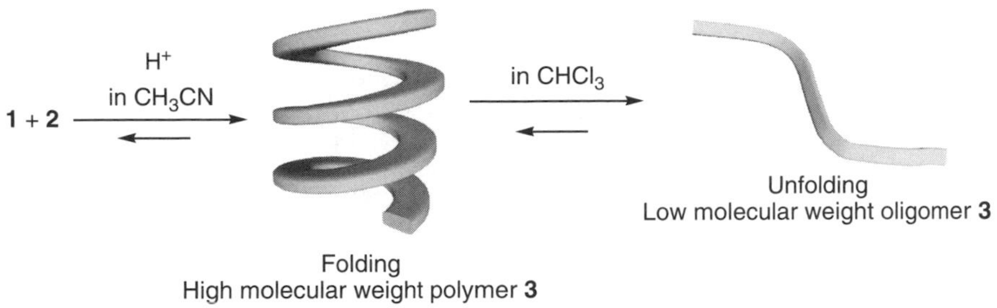

Figure 8.2 Schematic representation of the imine metathesis polymerization of **1** and **2** driven by folding of resulting polymer **3** [17,18].

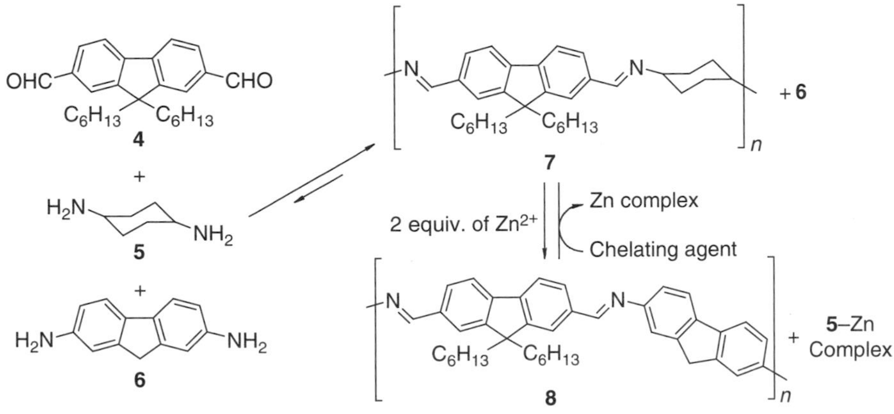

Scheme 8.2 Synthesis of changeable and tunable polymers **7** and **8** form the 1:1:1 mixture of **4:5:6** and effects of additions of Zn^{2+} ion and its chelating agent [20,21].

diamines. Furthermore, the addition of 2 equiv. of Zn^{2+} ion to the system promoted the generation of the fluorene-based polymer **8** because of the preferential formation of the complex of **5** with zinc ion. Then, the addition of a zinc ion chelating molecule (hexamethylhexacyclene) to the system reformed polymer **7**. The composition of polymers was fully reversible and controlled by modulation of the equilibrium by means of the additive and the nucleophilicities of the monomer units.

Dynamic polymers having C=N bonds responding to changes in a neat/solution environment were reported by Lehn et al. [22]. Polymers **9–12** were synthesized by the corresponding dialdehydes and diamines in the presence of anhydrous Na_2SO_4 (Fig. 8.3). The isotropic transition temperatures of polymers **9** and **10** were obtained at 72.5 and 68.4°C, respectively. Polymer **11** was not a film or solid but viscous oil with an isotropic phase at less than 40°C. In contrast, polymer **12** showed the highest isotropic transition temperature at about 160°C due to the existence of

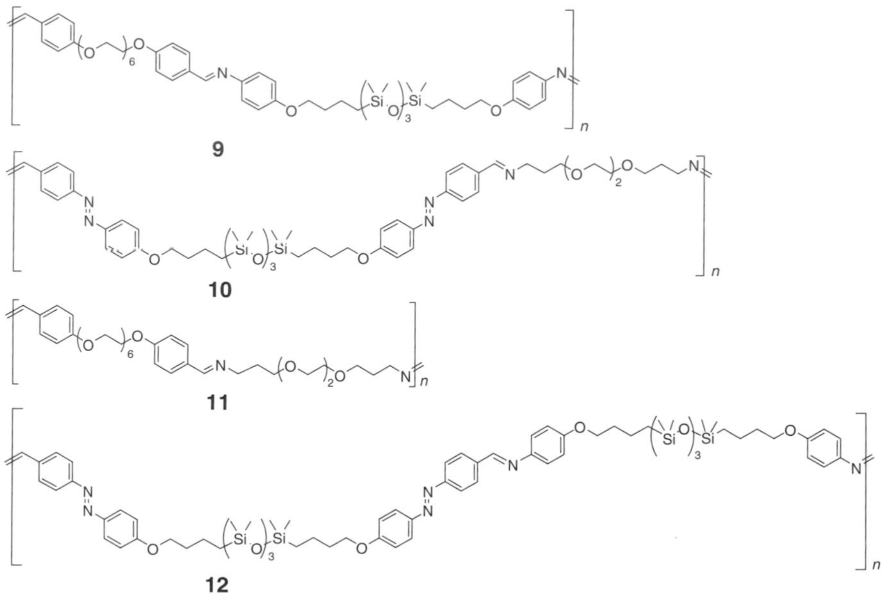

Figure 8.3 Dynamic polymers responding to changes in neat/solution environment [22].

strong mesogens. When equimolar amounts of polymers **9** and **10** were blended in chloroform with a catalytic amount of pentadecafluorooctanoic acid, these polymers were recombined by the C=N exchange reaction, giving a random copolymer with four components **9–12**. The composition of **9:10:11:12** after achieving equilibrium was approximately 3:3:2:2, respectively. Interestingly, in the neat condition the ratio of the four domains was quite different from that in solution state. The composition of **9:10:11:12** in the neat condition was 0:0:5:5, showing only the existence of domains of **11** and **12**. Furthermore, the successive solution–neat cycles by dissolution–evaporation operation displayed reversible switching of the composition of the four domains.

C=N bonds in acylhydrazone groups, which are formed by the condensation of hydrazides with carbonyl groups, exhibit reversibility under mild conditions. Like polyimines, polymers with acylhydrazone functionalities exhibit dynamic aspects through the reversibility of the azomethine bond. Skene and Lehn have reported the synthesis of polyacylhydrazones and their dynamic features (Scheme 8.3) [23]. High molecular weight polyacylhydrazone **13** was prepared by the condensation of the corresponding dihydrazide and dialdehyde in the presence of an acid catalyst. When **13** was treated with aryl dialdehyde **14** or aryl dihydrazide **15** in

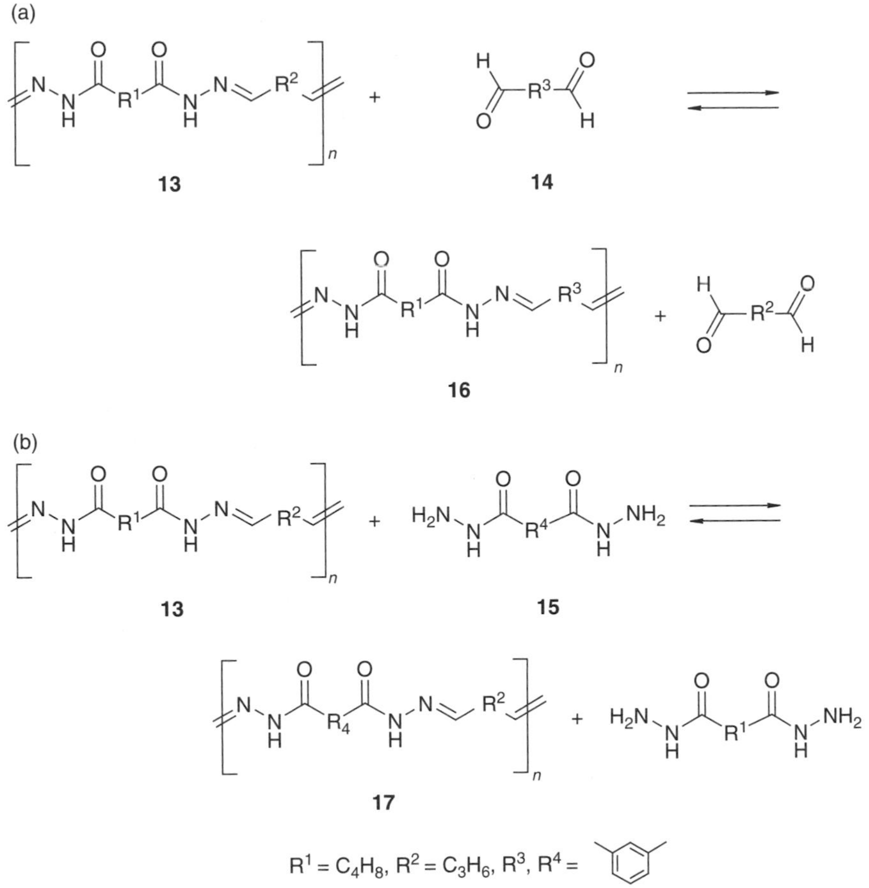

(a)

(b)

$$R^1 = C_4H_8, R^2 = C_3H_6, R^3, R^4 = $$

Scheme 8.3   Exchanges of the components in polyacylhydrazone **13** with (**a**) dialdehyde **14** and (**b**) bifunctional acylhydrazide **15** through the C=N exchange reaction [23].

DMSO in the absence of acid catalyst, the component exchanges were implemented by heating, and the exchange ratios were reached at approximately 50% with a limited heating time. Although the resulting polymers **16** and **17** did not reach equilibria, the component exchange based on the equilibrium reaction is significant for modification of the chemical properties of the resulting polymers.

Lehn's group has also reported a series of research related to the improvement of polymer properties using the dynamic feature of polyacylhydrazone (Figs. 8.4–8.7) [24,25]. Polymers containing acylhydrazone bonds in linear main chains, **25** and **26**, were prepared from the corresponding bifunctional acylhydrazides **18** and **19** and dialdehydes **21** and **22**.

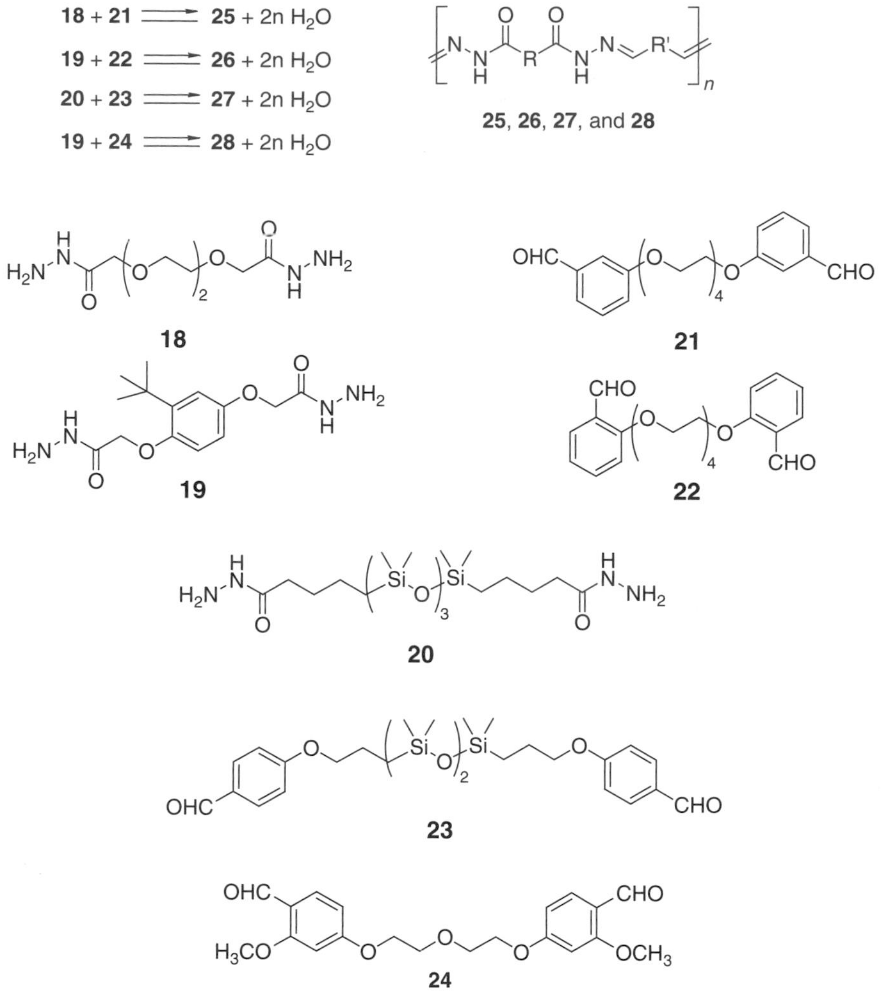

18 + 21 $\rightleftharpoons$ 25 + 2n H₂O

19 + 22 $\rightleftharpoons$ 26 + 2n H₂O

20 + 23 $\rightleftharpoons$ 27 + 2n H₂O

19 + 24 $\rightleftharpoons$ 28 + 2n H₂O

25, 26, 27, and **28**

18

21

19

22

20

23

24

Figure 8.4 Synthetic schemes of polyacylhydrazones **25–28** and structures of their monomers bifunctional acylhydrazides **18–20** and dialdehydes **21–24** [24,25].

Transparent film of the mixture of **25** and **26**

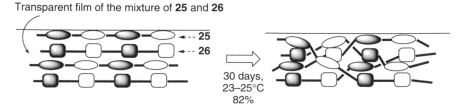

30 days, 23–25°C 82%

Figure 8.5 Schematic diagram of the crossover of components of blended polymers **25** and **26** through reversible and interchain C=N exchange reaction [24].

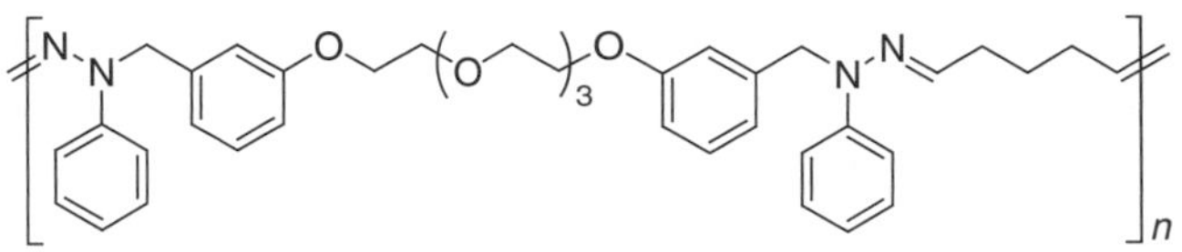

29   Colorless, slightly turbid film

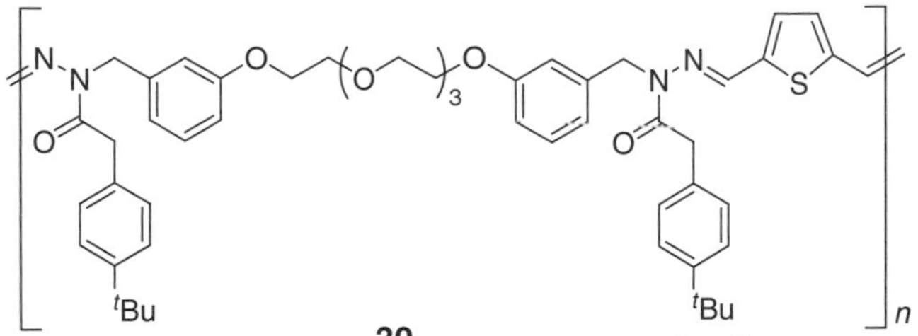

30   Light yellow film

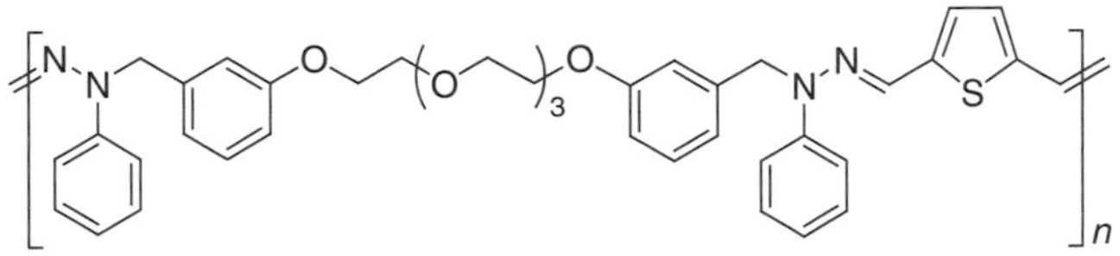

31   Dark orange film

Figure 8.6   Structures of polyacylhydrazones **29–31** [26].

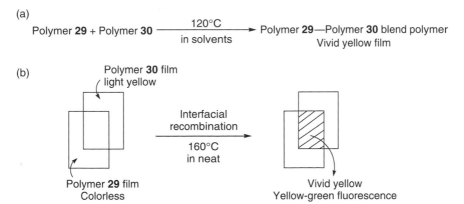

(a)

Polymer **29** + Polymer **30** $\xrightarrow[\text{in solvents}]{120°C}$ Polymer **29**—Polymer **30** blend polymer
Vivid yellow film

(b)

Polymer **30** film
light yellow

Interfacial
recombination

160°C
in neat

Polymer **29** film
Colorless

Vivid yellow
Yellow-green fluorescence

Figure 8.7   Schematic representation of the recombinations of polymer **29** and **30** to obtain visual color changes (**a**) in solution and (**b**) in neat [26].

They were blended with an acid catalyst in solution, and the film was prepared by evaporation of the solution. As time advanced, acylhydrazide bond exchange took place in the neat blend (Fig. 8.5). The segmental scale exchange reaction took place, and the resulting polymeric molecules were identical to the copolymer consisting of all four components **18, 19, 21**, and **22**. Change of the mechanical properties of dynamic polyacyl-hydrazones through component exchanges was investigated. Polymer **27**, having flexible siloxane-derived spacers, and polymer **28**, with rigid spacer groups, were synthesized in the absence of acid catalyst (Fig. 8.4). Viscoelasticity measurements revealed that Poly-**27** possessed rubber elasticity at room temperature. Polymer **28** prepared from the monomers with hard components **19** and **24** was not strong enough for the viscoelasticity measurements. In contrast, the film prepared by the blending and recombination of polymer **27** with monomers **19** and **24** in the presence of acids had a typical glass state up to around 50°C and rubber elasticity at around 70°C (Table 8.1). This result is due to blending at the molecular level by acyl-hydrazone bond exchanges. Interestingly, the film of a blend of polymers **27** and **28**, which mixed in the absence of acid catalyst (not recombined through the acylhydrazone bond exchanges), possessed mechanical properties differing from that of polymer **27** recombined with monomers **19** and **24**, because polymer **28** was microdispersed in a polymer **27** matrix. Thus, recombination plays a key role in the evolution of mechanical properties. The same group has also reported the chromogenic effect of the films consisting of dynamic polyacylhydrazones of polymers **29, 30**, and **31** based on the chromophore formation through bond exchanges (Fig. 8.6) [26]. Polymers **30** and **31** were obtained as films light yellow and dark orange in

Table 8.1 Mechanical properties of the exchangeable acylhydrazone polymer films

Film	T_g (°C)[a]	E'(GPa)[b]
27	10	0.015
28	100[c]	— [d]
27 recombined with **19** and **24**[e]	56	1.1
27 blended with **28**[f]	30	0.59

[a]Glass transition temperature determined by the loss elastic modulus E''.
[b]Storage elastic modulus E' at 25°C.
[c]Glass transition temperature determined by DSC.
[d]Not determined.
[e]Recombined through the imine exchange reaction.
[f]Not recombined through the imine exchange reaction but just blended.
Source: Reference 25.

color, respectively, although the film of polymer **29** was colorless. Mixing of polymers **29** and **30** in solution in the presence of acid led to a change in UV-visible absorption and fluorescence induced by bond recombination. The color of the film mixed in solution changed from light yellow to vivid yellow on heating to 120°C (Fig. 8.7). This color change originated from generation of the domain of polymer **31**, the film of which was light yellow in color. Under a neat condition, color and fluorescence changes of the film were observed at the interface of polymer **29** film and polymer **30** film. In Fig. 8.7b, polymer **29** film was overlapped with polymer **30** film, and the two were heated at 160°C. As a result, the color was changed at the overlapping area together with the generation of fluorescence.

8.2.2. Alkene Metathesis

Alkene metathesis is an essential method for the formation of carbon–carbon double bonds through a reversible process using organometallic catalysts, as represented by Grubbs catalyst [$(PCy_3)_2Ru(CHPh)Cl_2$], under mild conditions (Fig. 8.8) [27]. Ring-closing metathesis (RCM) has played a significant role in medium-sized ring synthesis and natural product syntheses [28–32]. Acyclic diene metathesis (ADMET), irreversibly leading to the linear polymer by the efficient volatilization of ethylene, is of significance in polymer syntheses. Cyclic polyolefins generated by RCM and linear polyolefins synthesized by ADMET can readily participate in ring-chain equilibria by successive metathesis of catalyst in closed systems. Therefore, metathesis operating under polymerization–cyclodepolymerization equilibrium has led to the production of thermodynamically controlled polymers. Grubbs et al. have reported the lithium ion-mediated thermodynamic control of unsaturated crown ether analogs (Scheme 8.4) [33]. Cyclic olefin **32** was prepared by RCM of the

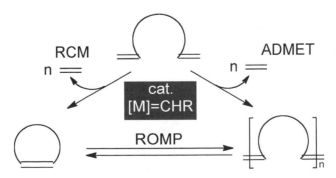

Figure 8.8   A manifold of alkene metathesis.

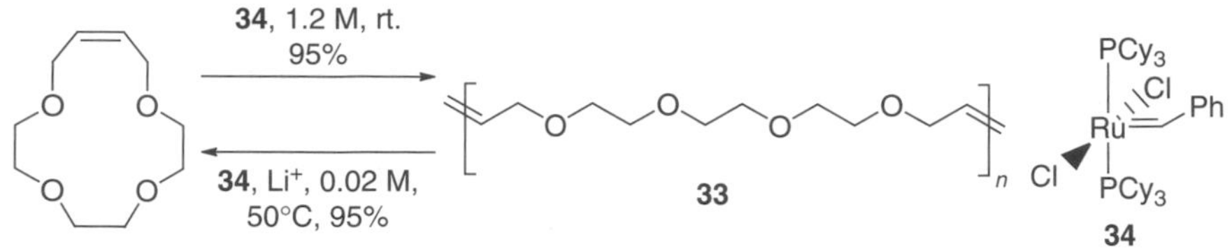

Scheme 8.4   A ring-opening metathesis of **32** and template-directed ring-closing metathesis of **33** catalyzed by **34** [33].

corresponding diene in the presence of catalyst **34** and Li^+ as a template at relatively high dilution method at 50°C. Ring-opening metathesis polymerization (ROMP) of **32** initiated by **34** proceeded smoothly with quantitative conversion to **33**. Cyclodepolymerization of **33** also performed under the RCM condition in the presence of Li^+ resulted in nearly quantitative conversion into **32**. Thus, the reversible formation and scission of a carbon–carbon double bond permits the production of thermodynamic polymeric materials that are tolerant of both acids and bases and that can be manipulated under milder conditions.

8.2.3. Alkyne Metathesis

Analogous to alkene and imine metathesis, alkyne metathesis can be a reversible process with an equilibrium constant close to unity. Recently, transition metal complexes that catalyze the alkyne metathesis of highly functionalized substrates under mild conditions were developed and applied to the synthesis of cyclic alkynes [34]. Zhang and Moore have reported the reversible feature of phenylene ethynylene derivatives using molybdenum-based metathesis catalyst (Scheme 8.5) [35]. Linear phenylene–ethynylene polymer **35** synthesized via Sonogashira crosscoupling of the corresponding diiodide and diacetylene was conducted via alkyne metathesis in a closed system at 30°C. After 22 hours, the starting polymers were converted to shorter oligomers with hexacycle **36** as the major product. Macrocycle **36** was reconverted to the oligomer with linear oligomeric **37** by the crossmetathesis reaction with diphenylacetylene. In addition, when monomeric **37** ($n = 1$) was subjected to alkyne metathesis in a closed system, the resulting product was identical to the product of the reaction of **36** and diphenylacetylene. These attempts at clarifying the reaction pathway have revealed that alkyne metathesis progresses via a thermodynamic process and becomes a powerful candidate for generating thermodynamically controlled conjugated polymer materials.

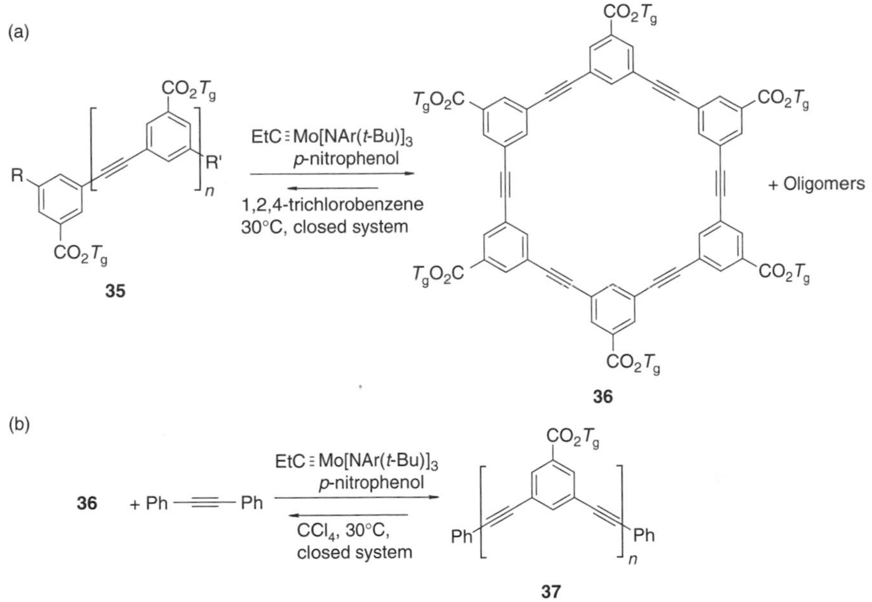

(a)

EtC ≡ Mo[NAr(t-Bu)]₃
p-nitrophenol

1,2,4-trichlorobenzene
30°C, closed system

CO_2T_g

35

CO_2T_g

+ Oligomers

36

(b)

36   + Ph ———— Ph

EtC ≡ Mo[NAr(t-Bu)]₃
p-nitrophenol

CCl₄, 30°C,
closed system

CO_2T_g

Ph   Ph

37

Scheme 8.5   Alkyne metathesis reactions of polymer **35** prepared from the (**a**) Sonogashira condition and (**b**) hexacycle **36** and diphenylacetylene [35].

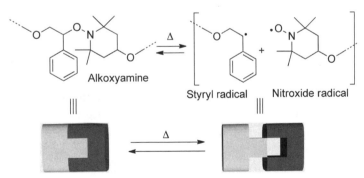

Alkoxyamine

Δ

Styryl radical   Nitroxide radical

III   III

Δ

Figure 8.9   Radical exchange reaction of alkoxyamine derivatives.

8.2.4. Radical Exchange Reaction

2,2,6,6-Tetramethylpiperidine-1-oxy (TEMPO)-containing alkoxyamine derivatives are widely used as unimolecular initiators for living radical polymerization [5]. The key step of the presently accepted mechanism of polymerization is the reversible capping of the polymer chain by the nitroxide radical. In 2002, Otsuka and Takahara applied the reversible

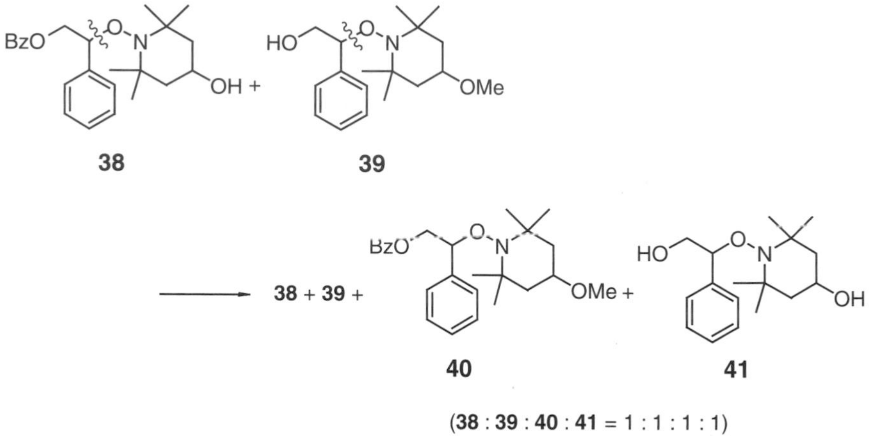

$(38 : 39 : 40 : 41 = 1 : 1 : 1 : 1)$

Scheme 8.6   The radical crossover reaction between alkoxyamine derivatives **38** and **39** [36].

nature of alkoxyamine to the synthesis of thermally reversible covalent polymer [36]. They used dynamic covalent units containing the TEMPO structure (Fig. 8.9). These units can generate styryl radicals and nitroxide radicals by heating, and the radical crossover reactions were continually carried out during heating. They first investigated the degree of exchange between alkoxyamine units using model compounds. Equimolar amounts of **38** and **39** were mixed in anisole, which was sealed under vacuum and heated at various temperatures (Scheme 8.6). As time advances, compounds **40** and **41** were yielded by the interchange reaction of **38** and **39**, and the molar ratios of **38, 39, 40**, and **41** were almost the same after achieving equilibrium. These studies have proved that alkoxyamines interchange above 60°C without yielding any by-products, and that the crossover reaction achieves equilibrium at 100°C after 12 hours. Based on the model reaction, dynamic polymer **43** with a thermally reversible alkoxyamine functionality of TEMPO was synthesized by polycondensation from TEMPO-diol **41** and adipoyl chloride (Fig. 8.10). To investigate the interchain crossover reaction, fractionated poly(alkoxyamine ester)s containing TEMPO having different molecular weights **43a** (M_n = 12000, M_w/M_n = 1.21) and **43b** (M_n = 4300, M_w/M_n = 1.17) were mixed and heated in a closed system. As time proceeded, peaks of the gel permeation chromatography (GPC) profile derived from **43a** and **43b** clearly fused into a unimodal peak of **43c** (M_n = 5600, M_w/M_n = 1.86). It strongly indicated that **43** is a dynamic polymer that can dissociate and associate reversibly in the polymer backbone triggered by heating.

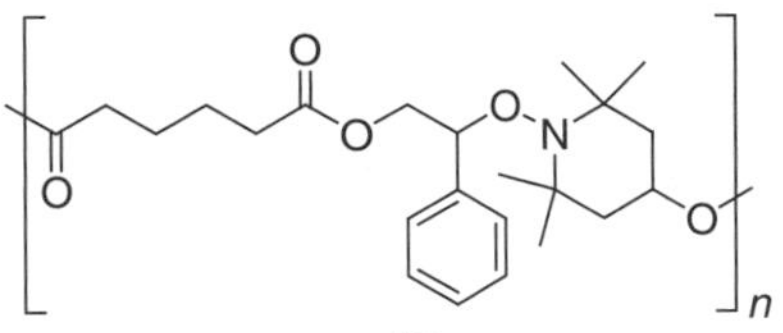

43

43a ($M_n = 12000$, $M_w/M_n = 1.21$)

+ $\xrightarrow{\Delta}$ 43c ($M_n = 5600$, $M_w/M_n = 1.86$)

43b ($M_n = 4300$, $M_w/M_n = 1.17$)

Figure 8.10   Structure of polyester bearing the alkoxyamine unit and change on molecular weight and molecular weight distribution of polymer **43** in the radical crossover reaction of **43a** and **43b** [36].

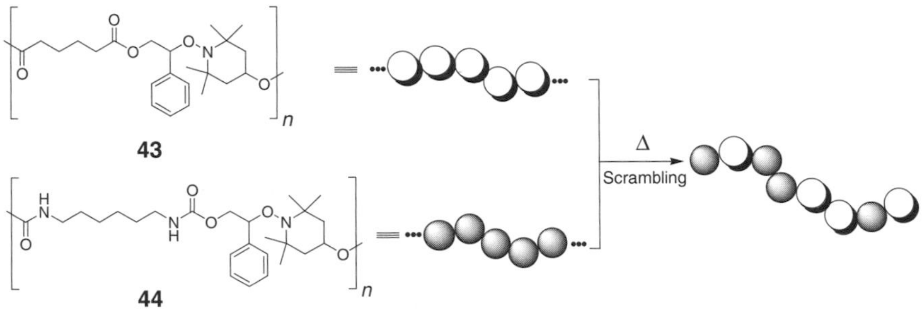

Figure 8.11   Schematic diagram of the "polymer scrambling" by the radical crossover reaction of polyester **43** and polyurethane **44** having alkoxyamine units [37].

The exchange in the alkoxyamine-based polymer occurs in a radical process that is tolerant of many functional groups. The exchange process is therefore applicable to polymers with various functional groups. TEMPO-based polyester **43** and polyurethane **44** were synthesized for studies of the scrambling of disparate polymers under thermodynamic control (Fig. 8.11) [37]. Two kinds of TEMPO-based polymers were mixed and heated in a closed system. After 24 hours when the crossover reaction achieved equilibrium, GPC and NMR analyses revealed that they were totally scrambled through bond recombination on the TEMPO units.

From the perspective of the basic mechanism of the radical exchange reaction of TEMPO-based exchangeable and tunable polymer, thermal reorganization behavior was studied by using polymer **43** [38]. The change in molecular weight during the radical crossover reaction between fractionated narrowly dispersed **43** with various molecular weights is shown in Table 8.2. In all cases, the M_n values of the postpolymers were midway

Table 8.2   Changes in molecular weights and molecular weight distributions in radical crossover reactions of **43**

Run	Prepolymer **43** (High M_w)		Prepolymer **43** (Low M_w)		Recombinated polymer[a]	
	M_n	M_w/M_n	M_n	M_w/M_n	M_n	M_w/M_n
1	17400	1.37	8200	1.18	9800	1.88
2	17400	1.37	6300	1.31	8400	1.77
3	17400	1.37	4300	1.17	6000	1.82
4	12000	1.21	6300	1.31	6900	1.71
5	12000	1.21	4300	1.17	5600	1.86

[a]Equimolar amounts of **43** with different M_w were mixed and heated at 100°C for 12 hours.
Source: Reference 38.

between the high and low M_n values of the prepolymers. These results suggested that polymers with various molecular weights were combinatorially formed by reorganization between polymers with different M_n values (Fig. 8.12). In addition, the chain transfer reaction of TEMPO-based polymer **43** was carried out by the addition of nitroxide stable free radical **45** in the reorganization process (Scheme 8.7). The competitive chain transfer reaction of the added free radical to the generated styryl radical caused a scission of the polymer chain (Table 8.3). These studies indicated that the macromolecular crossover reaction in TEMPO-based polymers proceeds by a radical process at moderately high temperature.

Based on these results, ring-opening polymerization of a TEMPO-based macrocycle **46a** was conducted by means of the radical crossover reaction of TEMPO units (Scheme 8.8) [39]. Macrocyclic ester **46a** containing TEMPO units was synthesized under a high-dilution condition from the corresponding TEMPO-diol **41** and adipoyl chloride. At high concentrations (7.2 wt%), the macrocycle **46a** was polymerized via intermolecular radical crossover reaction at 125°C. GPC and NMR analyses indicated that the polymerization progressed without any by-products and yielded high molecular weight polymer **46b**. The depolymerization of **46b** was allowed to proceed under the high dilution condition (0.40 wt%). These reversible features were controlled by only physical stimuli like concentrations and heating.

The radical exchange reaction of alkoxyamines has some advantages for the design of dynamic polymers. The reaction is controlled by temperature and affords the exchanged alkoxyamine units without by-products. The reaction can be applicable to polymers having various functional groups

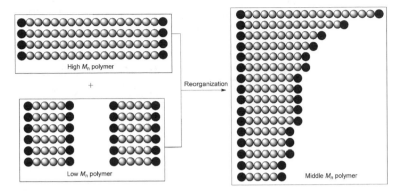

Figure 8.12   Combinatorial production by reorganization between high and low molecular weight polymers **43** [38].

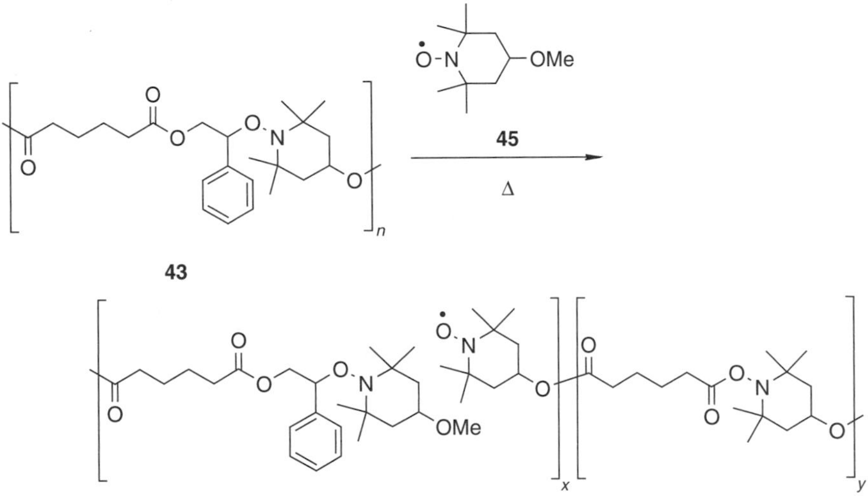

Scheme 8.7   Chain transfer reaction of polyester containing alkoxyamine units **43** caused by free radical **45** [38].

because the recombination process occurs through a radical process that is tolerant of many functional groups. Furthermore, nitroxide radical and styryl radical that are generated by homocleavage of the corresponding alkoxyamine completely recombined without homocoupling of styryl radicals. The cleavage and complementary recombination of alkoxyamine is one of the significant features of this reversible bond.

Table 8.3   Chain transfer reactions of polymer **43** by 4-methoxy-TEMPO free radical **45**

Run	Added **45** (equiv.)[a]	Time (hour)	Temperature (°C)	M_n	M_w/M_n
1	1.0	1.0	30	11800	1.21
2	1.0	0.5	100	8800	1.49
3	1.0	1.0	100	5100	1.45
4	1.0	3.0	100	1700	1.88
5	1.0	6.0	100	1200	1.71

[a]M_n and M_w/M_n of polymer **43** was estimated to 11,800 and 1.21, respectively.
Source: Reference 38.

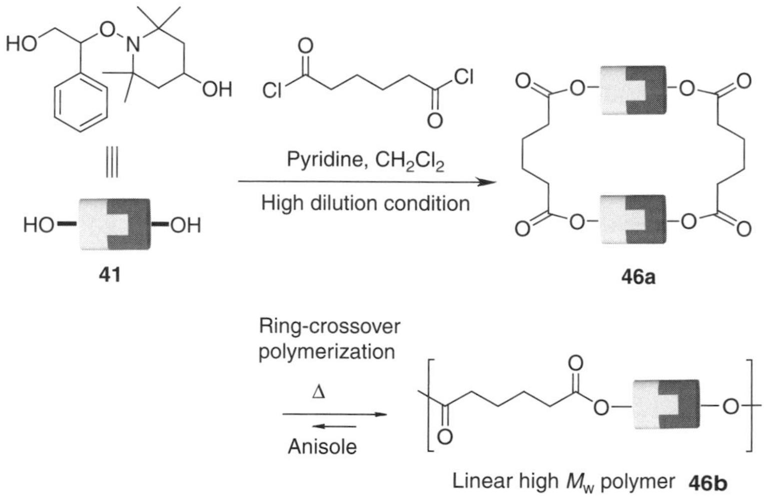

Scheme 8.8   Preparation of macrocycle **46a** having alkoxyamine units and ring-crossover polymerization of **46a** [39].

Dynamic formation of graft polymers was synthesized by means of the radical crossover reaction of alkoxyamines by using the complementarity between nitroxide radical and styryl radical (Fig. 8.13) [40]. Copolymer **48** having alkoxyamine units on its side chain was synthesized via atom transfer radical polymerization (ATRP) of TEMPO-based alkoxyamine monomer **47** and MMA at 50°C (Scheme 8.9). The TEMPO-based alkoxyamine-terminated polystyrene **49** was prepared through the conventional nitroxide-mediated free radical polymerization (NMP) procedure [5,41]. The mixture of copolymers **48** and **49** was heated in anisole

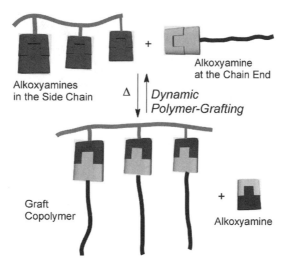

Figure 8.13 Schematic representation for the dynamic formation of graft polymer through radical crossover reactions of alkoxyamine units [32].

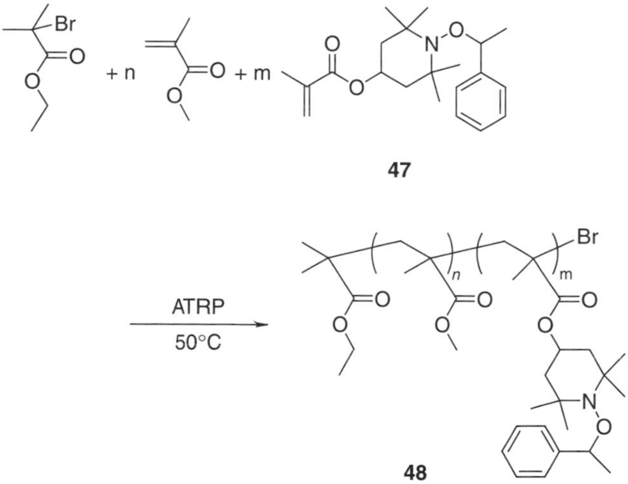

Scheme 8.9 Preparation of copolymer of **47** and MMA by atom transfer radical polymerization at 50°C [40].

at 100°C to implement the thermodynamic exchange reaction between the nitroxide radical on the side chains of **48** and styryl radical on the terminal of **49** (Scheme 8.10). SEC profiles derived from **48** clearly shifted to the higher molecular weight region with increasing reaction time. ^{1}H-NMR revealed

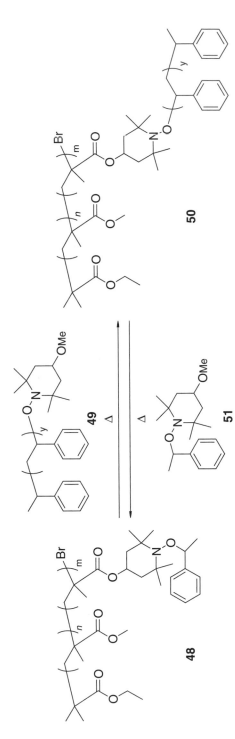

Scheme 8.10 Dynamic formation of graft polymer **50** prepared form copolymer **48** and TEMPO-based alkoxyamine-terminated polystyrene **49** [40].

that the degree of grafting was 58% when the excess of **49** was added to the reaction system. To demonstrate the consideration about dynamic nature of graft polymer **50**, the reversibility of the reaction system was also investigated. The mixture of **50** and an excess amount of nonfunctionalized alkoxyamine derivative **51** was heated in anisole at 100°C. According to LeChatelier's principle, **48** was reformed with **49**.

Moreover, the thermodynamic polymer crosslinking system based on the exchange reaction of alkoxyamine units has also been reported (Scheme 8.11) [42]. Two kinds of copolymers (**52** and **53**), having the potential to generate nitroxide radical and styryl radical, respectively, were synthesized from the methacrylic esters containing TEMPO units and MMA by the ATRP method at 50°C. The mixture of **52** and **53** was heated at various concentrations in anisole at 100°C to conduct the thermodynamic exchange reaction between the nitroxide radical on the side chain of **52** and styryl radical on the terminal of **53**. After heating, the solution became a gel at high concentrations. Furthermore, the reversibility of the reaction system was also investigated. The mixture of **54** and an excess amount of nonfunctionalized alkoxyamine derivative **51** was heated in anisole at 100°C. After heating, copolymers **52** and **53** were reformed, and there were no remarkable shoulder peaks in the high molecular weight region. These results indicate that the crosslinked points in **54** are thermally dissociable and that the reaction system is apparently reversible.

More recently, the programmed formation of star-like nanogels by using a radical crossover reaction of diblock copolymers with complementarily reactive alkoxyamine in side chains was reported (Fig. 8.14) [43,44]. Block copolymers consisting of PMMA blocks and methacrylic ester polymer blocks with TEMPO moieties were prepared by a two-step procedure (Scheme 8.12). Initially the PMMA blocks with a bromine atom at the terminal end were prepared by the ATRP method. Subsequently, random copolymerizations of MMA:**55**:**56** mixtures (5:1:1 and 20:1:1) were carried out using PMMA prepolymers by ATRP. The radical crossover reactions of diblock copolymers **57** were carried out at 100°C. Interestingly no macroscopic gelation of the system was observed even at high concentrations such as 10 wt% since the crosslinking on a macroscale was avoided by the existence of PMMA block. Furthermore, the morphology of crosslinking polymers with a gel part and branching chains were directly observed by scanning force microscopy (SFM). These results indicated the formation of thermodynamically controlled star-like nanogels by the radical crossover reaction of **57**. In addition, the mechanism of the formation of star-like nanogels **57** was also

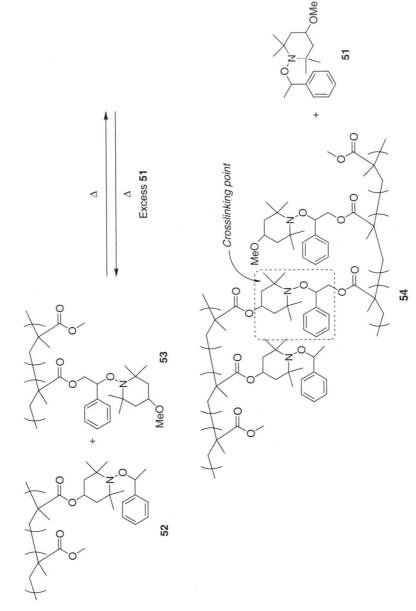

Scheme 8.11 Thermodynamic formation of crosslinked polymer **54** via radical crossover reaction of alkoxyamines in copolymers **52** and **53** [42].

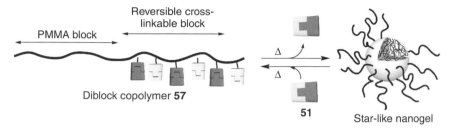

Figure 8.14 Schematic diagram of the interconversion between diblock copolymer and star-like nanogel through the radical crossover reaction of alkoxyamine units [44].

proposed based on the results of the time-dependent change of M_w and the radius of gyration estimated by GPC-multiangle laser light scattering (GPC-MALLS) and small-angle X-ray scattering (SAXS) measurements. First the intermolecular crosslinking reaction occurred to yield the primitive star-like nanogel. In the next stage the radical crossover reaction was prevented by the steric repulsion of PMMA block; the intramolecular crosslinking reaction, however, was carried out in the core of the star-like nanogel. In the final stage a thermodynamically stable structure was organized by the successive radical exchange reaction of alkoxyamine units on side chains. The correlation between the structures of diblock copolymers **57** and the core sizes and number of arms of the resulting star-like nanogels are listed in Table 8.4. Although a size difference in core parts between nanogels of **57a** and **57b** was not practically observed, the core size of nanogel **57c** was observed to be smaller due to the shorter length of the crosslinkable block. Similarly a lower height and a larger molecular weight were observed in nanogel **57d** due to the low crosslinking density. Thus, the molecular weight and the morphology of crosslinking **57** are clearly thermodynamically controlled by the relative proportion between PMMA block and crosslinkable TEMPO-containing block and by the composition of TEMPO-containing monomers in the second block through the selection in the equilibrium. In addition, the reversibility of the reaction system was also investigated. The mixture of the nanogel and the excess amount of nonfunctionalized alkoxyamine derivative **51** was heated in anisole at 100°C. After heating, the nanogel was reformed to the linear diblock copolymer **57**. These results proved that the nanogel was thermally dissociable and that the reaction system is apparently reversible.

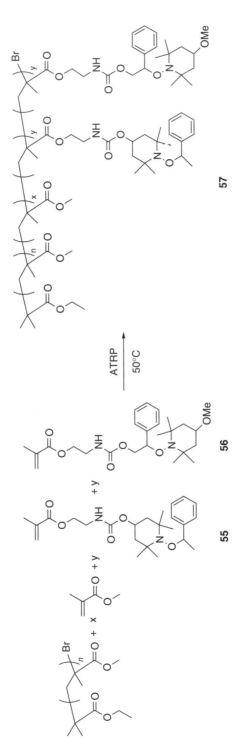

Scheme 8.12 Preparation of the diblock copolymer **57** by random copolymerization of PMMA block prepared by ATRP, methacrylic esters containing alkoxyamine units **55** and **56** [44].

Table 8.4 Correlation between compositions, molecular weights of diblock copolymer **57**, and core sizes and number of arms of star-like nanogels

Diblock copolymers	PMMA block M_n $(M_w/M_n)^a$	Block copolymer M_n $(M_w/M_n)^a$	x/y^b	Nanogels M_n^c	Core sizes (nm)^d	Number of arms^c
57a	23100	39000	5/1	1.1×10^6	2.15 ± 0.82	26.3
	(1.11)	(1.07)				
57b	55400	70000	5/1^c	1.3×10^6	2.43 ± 0.69	17.4
	(1.13)	(1.08)				
57c	23100	29700	5/1	3.5×10^5	1.02 ± 0.32	10.8
	(1.11)	(1.08)				
57d	22100	35400	20/1	2.4×10^6	1.00 ± 0.25	60.0
	(1.11)	(1.11)				

^{a}Estimated by SEC using PSt standards.
^{b}Relative proportion of MMA and TEMPO containing monomers in Scheme 8.10.
^{c}Determined by SEC-MALLS.
^{d}Measured using SFM height images.
Source: Reference 44.

8.2.5. Carbene Dimerization

In 2006 Kamplain and Bielawski proposed the existence of dynamic covalent polymers based on carbene dimerization (Scheme 8.13) [45]. They have used the equilibrium reaction of imidazol-2-ylidenes and their respective enetetraamine dimers [46]. Benzimidazole-based monomer **58** with a bifacially opposing carbene moiety prepared by the deprotonation of [5,5′]-bibenzimidazolium dibromide afforded high molecular weight polymeric material **59** by carbene dimerization. The resulting polymer was allowed to react with enetetraamine dimer **60** at 90°C via reversible carbene dimerization on polymer backbones. Furthermore, the reversibility of carbene dimerization in the main chain was no longer permitted by the insertion of transition metal and to produce the irreversible organometallic polymer **62**.

8.2.6. Other Equilibrium Reactions

Covalent polymers with reversible properties arising from dynamic covalent bonds such as disulfide exchange reaction [47–49], transesterification [50,51], transetherification [52], and boronate ester formation [53] were reported without respect to DCC. These studies should involve DCLs in

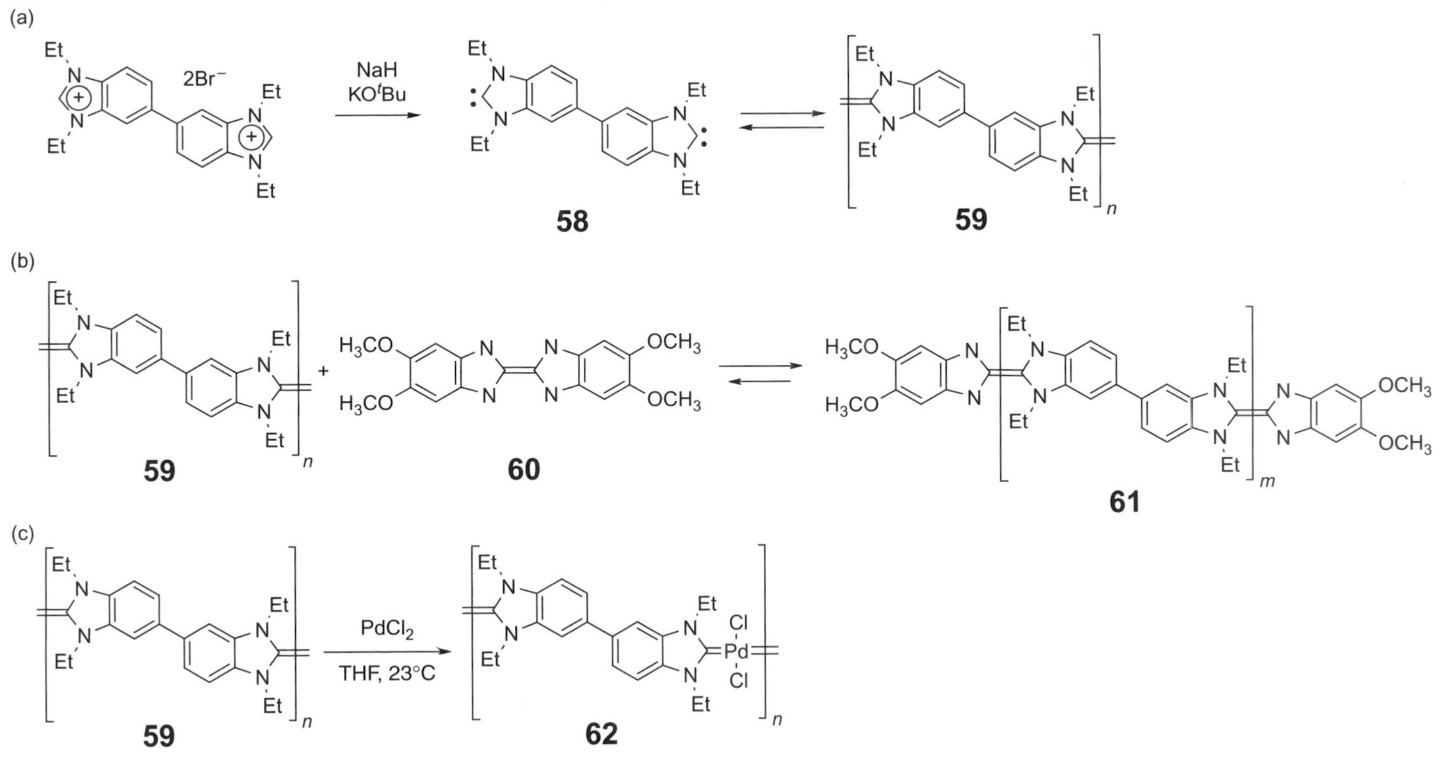

Scheme 8.13 Dynamic covalent polymers based on carbine dimerization. (a) Preparation of difunctional carbene **58** and polymerization of **58** via carbene dimerization; (b) Chain transfer reaction of **59** by the agency of monofunctional carbene **60**, and (c) Formation of the organometallic copolymer **62** by the insertion of PdCl$_2$ [45].

their preparation processes. They have potential to develop changeable and tunable materials produced through dynamic combinatorial approach.

8.3. Polymer-Supported Dynamic Combinatorial Chemistry

DCC is known to be an advanced technique for target-driven selection of high affinity ligands from the DCL. Recently, resin-bound DCC has been reported by Miller's group to simplify the identification (Fig. 8.15) [54,55]. They focused on the selection of DNA-binding compound M_x-M_x consisted of thiol derivative of M_x-SH. Nine library members of M_x-SH bearing different sets of amino acid residues were individually immobilized on the Tentagel-S resin and treated with thiopropanol. All members of M_x-SH and fluorescently labeled DNA were then added to the nine reaction vessels, and they were left for 24 hours. After draining and washing, resins were imaged via fluorescence microscopy. For example, when the intercalator M_5 is specifically bound to the target DNA, a resin-immobilized M_5 only exhibits fluorescence because of the existence of M_5-M_5. This methodology is applicable in screening large libraries or complex libraries in which intercalators consisting of different M_x-M_y components are also bound to DNA together with M_x-M_x or M_y-M_y.

8.4. Summary and Prospects

Dynamic combinatorial chemistry for searching for the most stable compounds has evolved in the areas of preferential syntheses of macrocycles and interlocked molecules with the evolution of supramolecular chemistry and dynamic covalent chemistry. Unlike the existing methodology in polymer synthesis, which pursues syntheses of polymers with structural unities without producing diverse polymers having various compositions, molecular weights, and topologies, the dynamic combinatorial approach in polymer synthesis provides changeable and tunable polymeric materials under thermodynamic control, as described in this chapter. These polymeric materials having dynamic covalent bonds were generated via DCLs constructed by initiations of equilibrium reactions induced by catalysts, additives, concentrations, and temperature change. In conjunction with the supramolecular polymers [7,8] and reversible polymers using not equilibrium reactions but reversible reactions like the retro Diels–Alder reaction [56], this novel strategy related to DCC should be expected to contribute to the development of "smart materials" with advanced functions such as self-mending,

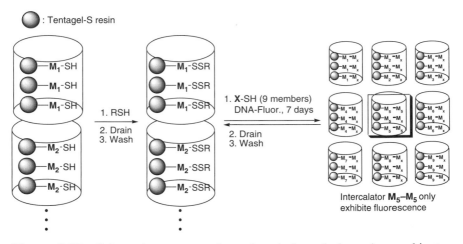

Figure 8.15 Schematic representation of resin-bound dynamic combinatorial chemistry and structures of the family of intercalating DNA binding agents M_{1-9} [54].

error-checking, and proof-reading abilities. At this time, a limited number of reports in this area have been published by some research groups. However, as polymer chemists will seek more complex and highly functionalized polymeric materials in the future, the dynamic combinatorial approach will become more and more spotlighted.

References

1. Matyjaszewski, G.; Gnanou, Y.; Leibler, L., editors. Macromolecular engineering. Volume 1: Synthetic technique. Wiley-VCH, Weinheim, **2007**.

2. Kamigaito, M.; Ando, T.; Sawamoto, M. Metal-catalyzed living radical polymerization. *Chem. Rev.* **2001**, *101*, 3689–3745.

3. Hadjichristidis, N.; Pitsikalis, M.; Pispas, S.; Iatrou, H. Polymers with complex architecture by living anionic polymerization. *Chem. Rev.* **2001**, *101*, 3747–3792.

4. Aoshima, S.; Yoshida, T.; Kanazawa, A.; Kanaoka, S. New stage in living cationic polymerization. *J. Polym. Sci. Part A Polym. Chem.* **2007**, *45*, 1801–1813.

5. Hawker, C. J.; Bosman, A. W.; Harth, E. New polymer synthesis by nitroxide mediated living radical polymerizations. *Chem. Rev.* **2001**, *101*, 3661–3688.

6. Yokozawa, T.; Yokoyama, A. Chain-growth polycondensation: The living polymerization process in polycondensation. *Prog. Polym. Sci.* **2007**, *32*, 147–172.

7. Brunsveld, L.; Folmer, B. J. B.; Meijer, E. W.; Sijbesma, R. P. Supramolecular polymers. *Chem. Rev.* **2001**, *101*, 4071–4097.

8. Kato, T.; Mizoshita, N.; Kanie, K. Hydrogen-bonded liquid crystalline materials: Supramolecular polymeric assembly and the induction of dynamic function. *Macromol. Rapid Commun.* **2001**, *22*, 797–814.

9. Lehn, J.-M. Dynamers: Dynamic molecular and supramolecular polymers. *Prog. Polym. Sci.* **2005**, *30*, 814–831.

10. Lehn, J.-M. From supramolecular chemistry towards constitutional dynamic chemistry and adaptive chemistry. *Chem. Soc. Rev.* **2007**, *36*, 151–160.

11. Huc, I.; Lehn, J.-M. Virtual combinatorial library: Dynamic generation of molecular and supramolecular diversity by self-assembly. *Proc. Natl. Acad. Sci. U.S.A.* **1997**, *94*, 2106–2110.

12. Lehn, J.-M. Dynamic combinatorial chemistry and virtual combinatorial libraries. *Chem. Eur. J.* **1999**, *5*, 2455–2463.

13. Furlan, R. L. E.; Otto, S.; Sanders, J. K. M. Supramolecular templating in thermodynamically controlled synthesis. *Proc. Natl. Acad. Sci. U.S.A.* **2002**, *99*, 4801–4804.

14. Corbett, P. T.; Leclaire, J.; Vial, L.; West, K. R.; Wietor, J.-L.; Sanders, J. K. M.; Otto, S. Dynamic combinatorial chemistry. *Chem. Rev.* **2006**, *106*, 3652–3711.

15. Ladame, S. Dynamic combinatorial chemistry: On the road to fulfilling the promise. *Org. Biomol. Chem.* **2008**, *6*, 219–226.

16. Rowan, S. J.; Cantrill, S. J.; Cousins, G. R. L.; Sanders, J. K. M.; Stoddart, J. F. Dynamic covalent chemistry. *Angew. Chem. Int. Ed.* **2002**, *41*, 898–952.

17. Zhao, D.; Moore, J. S. Reversible polymerization driven by folding. *J. Am. Chem. Soc.* **2002**, *124*, 9996–9997.

18. Zhao, D.; Moore, J. S. Folding-driven reversible polymerization of oligo(*m*-phenyleneethynylene) imine: Solvent and starter sequence studies. *Macromolecules* **2003**, *36*, 2712–2720.

19. Oh, K.; Jeong, K.-S.; Moore, J. S. Folding-driven synthesis of oligomers. *Nature* **2001**, *414*, 889–893.

20. Giuseppone, N.; Lehn, J.-M. Constitutional dynamic self-sensing in a zinc[II]/polyiminofluorenes system. *J. Am. Chem. Soc.* **2004**, *126*, 11448–11449.

21. Giuseppone, N.; Fuks, G.; Lehn, J.-M.Tunable fluorene-based dynamers through constitutional dynamic chemistry. *Chem. Eur. J.* **2006**, *12*, 1723–1735.

22. Chow, C.-F.; Fujii, S.; Lehn, J.-M. Crystallization-driven constitutional changes of dynamic polymers in response to neat/solution conditions. *Chem. Commun.* **2007**, 4363–4365.

23. Skene, W. G.; Lehn, J.-M. Dynamers: Polyacylhydrazone reversible covalent polymers, component exchange, and constitutional diversity. *Proc. Natl. Acad. Sci. U.S.A.* **2004**, *101*, 8270–8275.

24. Ono, T.; Nobori, T.; Lehn, J.-M. Dynamic polymer blends: Component recombination between neat dynamic covalent polymers at room temperature. *Chem. Commun.* **2005**, 1522–1524.

25. Ono, T.; Fujii, S.; Nobori, T.; Lehn, J.-M. Soft-to-hard transformation of the mechanical properties of dynamic covalent polymers through component incorporation. *Chem. Commum.* **2007**, 46–48.

26. Ono, T.; Fujii, S.; Nobori, T.; Lehn, J.-M. Optodynamers: Expression of color and fluorescence at the interface between two films of different dynamic polymers. *Chem. Commum.* **2007**, 4360–4362.

27. Grubbs, R. H. Olefin metathesis. *Tetrahedron* **2004**, *60*, 7117–7140.

28. Grubbs, R. H.; Chang, S. Recent advances in olefin metathesis and its application in organic synthesis. *Tetrahedron* **1998**, *54*, 4413–4450.

29. Conrad, J. C.; Eelman, M. D.; Duarte Silva, J. A.; Monfette, S.; Parnas, H. H.; Snelgrove, J. L.; Fogg, D. E. Oligomers as intermediates in ring-closing metathesis. *J. Am. Chem. Soc.* **2007**, *129*, 1024–1025.

30. Hodge, P.; Kamau, S. D. Entropically driven ring-opening-metathesis polymerization of macrocyclic olefins with 21–84 ring atoms. *Angew. Chem. Int. Ed.* **2003**, *42*, 2412–2414.

31. Kamau, S. D.; Hodge, P.; Hall, A. J.; Dad, S.; Ben-Haida, A. Cyclo-depolymerization of olefin-containing polymers to give macrocyclic oligomers by metathesis and the entropically-driven ROMP of the olefin-containing macocyclic esters. *Polymer* **2007**, *48*, 6808–6822.

32. Grasillas, A.; Pérez-Castells, J. Macrocyclization by ring-closing metathesis in the total synthesis of natural products: Reaction conditions and limitations. *Angew. Chem. Int. Ed.* **2006**, *45*, 6086–6101.

33. Marsella, M. J.; Maynard, H. D.; Grubbs, R. H. Template-directed ring-closing metathesis: Synthesis and polymerization of unsaturated crown ether analogs. *Angew. Chem. Int. Ed.* **1997**, *36*, 1101–1103.

34. Zhang, W.; Moore, J. S. Alkyne metathesis: Catalysts and synthetic applications. *Adv. Synth. Catal.* **2007**, *349*, 93–120.

35. Zhang, W.; Moore, J. S. Reaction pathways leading to arylene ethynylene macrocycles via alkyne metathesis. *J. Am. Chem. Soc.* **2005**, *127*, 11863–11870.

36. Otsuka, H.; Aotani, K.; Higaki, Y.; Takahara, A. A dynamic (reversible) covalent polymer: Radical crossover behavior of TEMPO-containing poly(alkoxyamine ester)s. *Chem. Commun.* **2002**, 2838–2839.

37. Otsuka, H.; Aotani, K.; Higaki, Y.; Takahara, A. Polymer scrambling: Macromolecular radical crossover reaction between the main chains of alkoxyamine-based dynamic covalent polymers. *J. Am. Chem. Soc.* **2003**, *125*, 4064–4065.

38. Otsuka, H.; Aotani, K.; Higaki, Y.; Amamoto, Y.; Takahara, A. Thermal reorganization and molecular weight control of dynamic covalent polymers containing alkoxyamines in their main chains. *Macromolecules* **2007**, *40*, 1429–1434.

39. Yamaguchi, G.; Higaki, Y.; Otsuka, H.; Takahara, A. Reversible radical ring-crossover polymerization of an alkoxyamine-containing dynamic covalent macrocycle. *Macromolecules* **2005**, *38*, 6316–6320.

40. Higaki, Y.; Otsuka, H.; Takahara, A. Dynamic formation of grafted polymers via radical crossover reaction of alkoxyamines. *Macromolecules* **2004**, *37*, 1696–1701.

41. Sciannamea, V.; Jérôme, R.; Detrembleur, C. In-situ nitroxide-mediated radical polymerization (NMP) processes: Their understanding and optimization. *Chem. Rev.* **2008**, *108*, 1104–1126.

42. Higaki, Y.; Otsuka, H.; Takahara, A. A thermodynamic polymer cross-linking system based on radically exchangeable covalent bonds. *Macromolecules* **2006**, *39*, 2121–2125.

43. Amamoto, Y.; Higaki, Y.; Matsuda, Y.; Otuska, H.; Takahara, A. Programmed formation of nanogels via a radical crossover reaction of complementarily reactive diblock copolymers. *Chem. Lett.* **2007**, *36*, 1098–1099.

44. Amamoto, Y.; Higaki, Y.; Matsuda, Y.; Otuska, H.; Takahara, A. Programmed thermodynamic formation and structure analysis of star-like nanogels with core cross-linked by thermally exchangeable dynamic covalent bonds. *J. Am. Chem. Soc.* **2007**, *129*, 13298–13304.

45. Kamplain, J.; Bielawski, C. W. Dynamic covalent polymers based upon carbene dimerization. *Chem. Commun.* **2006**, 1727–1729.

46. Liu, Y.; Lindner, P. E.; Lemal, D. M. Thermodynamics of a diaminocarbene–tetraaminoethylene equilibrium. *J. Am. Chem. Soc.* **1999**, *121*, 10626–10627.

47. Tsarevsky, N. V.; Matyjaszewski, K. Reversible redox cleavage/coupling of polystyrene with disulfide or thiol groups prepared by atom transfer radical polymerizaion. *Macromolecules* **2002**, *35*, 9009.

48. Endo, K.; Shiroi, T.; Murata, N.; Kojima, G.; Yamanaka, T. Synthesis and characterization of poly(1,2-dithiane). *Macromolecules* **2004**, *37*, 3143–3150.

49. Oku, T.; Furusho, Y.; Takata, T. A concept for recyclable cross-linked polymer: Topologically networked polyrotaxane capable of undergoing reversible assembly and disassembly. *Angew. Chem. Int. Ed.* **2004**, *43*, 966–969.

50. Kricheldorf, H. R. Macrocycles. 21. Role of ring–ring equilibria in thermodynamically controlled polycondensations. *Macromolecules* **2003**, *36*, 2302–2308.

51. Berkane, C.; Mezoul, G.; Lalot, T.; Brigodiot, M. Lipase-catalyzed polyester synthesis in organic medium. Study of ring–chain equilibrium. *Macromolecules* **1997**, *30*, 7729–7734.

52. Colquhoum, H. M.; Lewis, D. F.; Ben-Haida, A.; Hodge, P. Ring-chain interconversion in high-performance polymer system. 2. Ring-opening polmerization–copolyetherification in the synthesis of aromatic poly(ether sulfones). *Macromolecules* **2003**, *36*, 3775–3778.

53. Nakazawa, I.; Suda, S.; Masuda, M.; Asai, M.; Shimizu, T. pH-dependent reversible polymers formed from cyclic sugar- and aromatic boronic acid-based bolaamphiphiles. *Chem. Comun.* **2000**, 881–882.

54. McNaughton, B. R.; Miller, B. L. Resin-bound dynamic combinatorial chemistry. *Org. Lett.* **2006**, *8*, 1803–1806.

55. McNaughton, B. R.; Gareiss, P. C.; Miller, B. L. Identification of a selective small-molecule ligand for HIV-2 frameshift-inducing stem-loop RNA form an 11,325 member resin bound dynamic combinatorial library. *J. Am. Chem. Soc.* **2007**, *129*, 11306–11307.

56. Bergman, S. D.; Wudl, F. Mendable polymers. *J. Mater. Chem.* **2008**, *18*, 41–62.

Index